RADIO RESOURCE MANAGEMENT FOR MULTIMEDIA QoS SUPPORT IN WIRELESS NETWORKS

D1480267

Related Recent Titles

Video Content Analysis Using Multimodal Information: For Movie Content Extraction, Inde
and Representation
Ying Li and C.-C. Jay Kuo
ISBN 1-4020-7490-5, 2003
http://www.wkap.nl/prod/b/1-4020-7490-5

Semantic Video Object Segmentation for Content-Based Multimedia Applications
Ju Guo and C.-C. Jay Kuo
ISBN 0-7923-7513-0, 2001
http://www.wkap.nl/prod/b/0-7923-7513-0

Content-Based Audio Classification and Retrieval for Audiovisual Data Parsing
Tong Zhang and C.-C. Jay Kuo
ISBN 0-7923-7287-5, 2001
http://www.wkap.nl/prod/b/0-7923-7287-5

RADIO RESOURCE MANAGEMENT FOR MULTIMEDIA QoS SUPPORT IN WIRELESS NETWORKS

by

Huan Chen
National Chung Cheng University, Taiwan

Lei Huang
Loyola Marymount University, U.S.A.

Sunil Kumar
Clarkson University, U.S.A.

C.-C. Jay Kuo
University of Southern California, U.S.A.

KLUWER ACADEMIC PUBLISHERS
Boston / Dordrecht / New York / London

Distributors for North, Central and South America:
Kluwer Academic Publishers
101 Philip Drive
Assinippi Park
Norwell, Massachusetts 02061 USA
Telephone (781) 871-6600
Fax (781) 871-6528
E-Mail: <kluwer@wkap.com>

Distributors for all other countries:
Kluwer Academic Publishers Group
Post Office Box 322
3300 AH Dordrecht, THE NETHERLANDS
Telephone 31 78 6576 000
Fax 31 78 6576 254
E-Mail: <services@wkap.nl>

 Electronic Services <http://www.wkap.nl>

Library of Congress Cataloging-in-Publication Data

Radio Resource Management for Multimedia QoS Support in Wireless Networks
Huan Chen, Lei Huang, Sunil Kumar, C.-C. Jay Kuo
ISBN 1-4020-7623-1

CONTENTS

Preface

Due to the great success and enormous impact of IP networks, Internet access (such as sending and receiving e-mails) and web browsing have become the ruling paradigm for next generation wireless systems. On the other hand, great technological and commercial success of services and applications is being witnessed in mobile wireless communications with examples of cellular, PCS voice telephony and wireless LANs. The service paradigm has thus shifted from the conventional voice service to seamlessly integrated high quality multimedia transmission over broadband wireless mobile networks. The multimedia content may include data, voice, audio, image, video and so on. With availability of more powerful portable devices, such as PDA, portable computer and cellular phone, coupled with the easier access to the core network (using a mobile device), the number of mobile users and the demand for multimedia-based applications is increasing rapidly. As a result, there is an urgent need for a system that supports heterogeneous multimedia services and provides seamless access to the desired resources via wireless connections.

Therefore, the convergence of multimedia communication and wireless mobile networking technologies into the next generation wireless multimedia (WMM) networks with the vision of "anytime, anywhere, anyform" information system is the certain trend in the foreseeable future. However, successful combination of these two technologies presents many challenges such as available spectral bandwidth, energy efficiency, seamless end-to-end communication, robustness, security, etc. To satisfy the growing demand for new innovative broadband wireless multimedia services, the currently deployed 2G wireless systems are evolving towards the third-generation (3G) systems to offer more advanced services. The 3G systems can support multimedia traffic at a target transmission rate of up to 2 Mbps for static mobile users and 144 Kbps for high mobility users. The 3G wireless technology, known as IMT-2000 (International Mobile Telecommunications-2000), is a unified technology standard, to provide a high quality, worldwide roaming capability on a small portable terminal.

One key issue in providing multimedia services over a mobile wireless network is the quality of service (QoS) support in the presence of changing network connectivity due to user mobility and shared, noisy, highly variable and limited wireless communication links.

Different techniques have been proposed in the literature to support QoS for multimedia applications at different layers of wireless networks. At the application layer, new multimedia coding systems, e.g. MPEG-4 and H.263L video codec, are adaptive to the changing network conditions, such as bandwidth and link quality. At the network layer, the protocols that are capable of handling mobility management and seamless connectivity are being developed. Routing mechanisms are also extended to be QoS aware and able to handle mobility. At the data link layer, medium access control is modified so that reservations and QoS guarantees can be supported. Similarly, error correction mechanisms can protect against higher and varying error rates of wireless links. At the physical layer, modulation and power control schemes are also designed to be QoS aware.

In this research monograph, we discuss network layer mechanisms, i.e. the resource reservation estimation (RRE) and the call admission control (CAC), under a unified radio resource management (RRM) framework to support multimedia applications in the next generation wireless cellular networks, with various QoS requirements such as bandwidth, delay/jitter, and priority. CAC schemes enable the system to provide QoS to the new as well as existing calls. Resource reservation scheme, such as Guard Channel, is used to reserve the resources for certain high priority calls. On the other hand, network is required to take the advantage of resource sharing among traffic in order to achieve better channel utilization. Obtaining a right balance between the two opposing criteria is, however, a big challenge. Our schemes cover RRM design for the channel-based wireless system, such as the time division multiple access (TDMA) and the frequency division multiple access (FDMA) systems, as well as the interference-based code division multiple access (CDMA) system.

In a cellular wireless system, geographic region is divided into small sized cells. Unlike wired networks, communication entities in cellular networks change their connectivity via handoff when they move from one cell to another. The use of micro or pico-sized cells makes the role of handoff procedures very important in maintaining the service continuity and QoS guarantees to the multimedia applications. The solution to this challenge is the proper management of scarce resources to satisfy both the service provider and mobile users. Sometimes, these two goals may however be conflicting to each other. Our schemes also address the issue of how to provide seamless handoff to mobile users, under the constraint of limited resources.

The organization of this research monograph is as follows. A brief overview of related issues and previous work is given in Chapters 1 and 2. In Chapter 3, mathematical models are discussed to analyze connection-level QoS in wireless multimedia networks. It provides the fundamental understanding of connection-level QoS for different classes of applications. The basic modeling methodology are used for the analysis of more complex scenarios in the later chapters.

In Chapter 4, an adaptive resource management system for multimedia services in a wireless network environment is discussed. The objective is to provide the desired QoS to different applications, under the constraints of limited and varying network bandwidth resources. Based on the concepts of service model and application profile, resources can be allocated adaptively to each service class by employing adaptive admission control and resource reservation schemes. In Chapter 5, dynamic call admission control schemes based on the single threshold guard channel (GC) scheme are presented for CBR and VBR traffic, respectively.

Chapter 6 addresses the issue of how to provide seamless handoff to mobile users with multiple priority traffic classes in FDMA/TDMA system. Multiple-threshold fixed GC scheme and dynamic GC scheme using SNR and distance information in the neighboring cells are presented. The interaction among communication elements and other implementation issues are also discussed. In Chapter 7, an efficient radio resource management (RRM) for CDMA-based cellular communication systems using the interference guard margin (IGM) scheme is presented. The resulting CAC scheme gives preferential treatment to higher priority handoff calls by pre-reserving a certain amount of resource in terms of IGM.

The RRM and CAC schemes presented in Chapters 4 and 5 address both connection-level and packet-level QoS. However, the schemes presented in chapters 6 and 7 address connection-level QoS only, whereas with more considerations on traffic profiles in the real-world wireless systems, including mobile terminal's data rate, different levels of priorities, mobility and rate adaptivity characteristics.

In Chapter 8, we formulate the optimal RRM design as the MDP optimization problem. MDP can be used to derive an optimal call admission policy in a stationary sense. In most cases, where the traffic condition does not change rapidly, the MDP-based call admission policy provides a better trade-off between optimality and complexity. The CAC policy can be controlled by choosing appropriate actions to accept or reject calls of different classes according to the current system state. Finally, the summary of presented techniques and future research directions are presented in Chapter 9.

Note that different chapters were completed at different stages of time, and we did not make special attempt to unify notations and terminologies in all chapters so that it is best to treat them independently at each chapter.

Acknowledgements

Dr. Huan Chen would like to thank his family for their extreme love and support for his study. He would also like to thank the National Chung Cheng University (in Taiwan) for their support.

Dr. Lei Huang would like to thank her parents, Shaowu and Yingchao, for their tremendous help in taking care of her baby during the delivery of the book. She would also like to thank her husband, Renzhong, for his endless love and support.

Dr. Sunil Kumar would like to thank his wife (Rajni) and sons (Paras and Shubham) for their unconditional support and love. He would also like to thank Clarkson University for their support.

Dr. C.-C. Jay Kuo would like to thank the University of Southern California for providing a dynamic and stimulating research environment.

Finally we thank the editor and staff at Kluwer Academic Publishers for their efforts throughout the production process. Our special thanks also go to Linda Varilla for carrying out the task of proofreading in a rapid and supportive fashion.

Chapter 1

INTRODUCTION

1.1 Quality of Service Issues in Wireless Multimedia Networks

Multimedia systems have attracted great attention during the past few years. Many multimedia systems with advanced information technologies have been developed in a distributed fashion. To enable access to distributed multimedia data, multimedia communication (as a convergence of information and telecommunication technologies) has been an active research area, and has promoted new applications with real-time video and audio streaming. On the other hand, great technological and commercial success of services and applications is being witnessed in mobile wireless communications with examples of cellular, PCS voice telephony and wireless LANs. When portable computers become more powerful, and the accessibility of a fixed network from a mobile host becomes easier, the number of mobile users and the demand for multimedia information by mobile users will increase rapidly. The core issue of providing multimedia services over a mobile wireless network is the quality of service (QoS) support in the presence of changing network connectivity due to user mobility and shared, noisy, highly variable and limited wireless communication links. In this research, we address the problem of providing QoS to multimedia applications in the next generation wireless networks.

1.1.1 Multimedia Applications

According to different delay requirements, multimedia applications could be classified into real-time and non-real-time applications.

Real-time applications cannot tolerate large delay and delay variation. Most interactive video/audio applications, such as video conferencing and telephone calls, have a stringent timing constraint and, thus, belong to real-time applications. On the other hand, non-real-time applications can tolerate relatively large delay and delay variation. Most traditional data applications, such as Telnet, FTP, email, and so on, can work without guarantees of timely data delivery. These applications are also called elastic, since they are able to stretch gracefully when delay increases; although these applications can benefit from short delay. The delay requirements vary from interactive applications like Telnet, to more asynchronous ones like email, with interactive bulk transfers like FTP in the middle.

In terms of loss/error requirement, real-time applications (e.g., voice and video) can generally tolerate some loss/error. For example, most modern video compression techniques such as H.263+ [1] and MPEG-4 [2] have various error resilience options (e.g. reference picture selection (RPS) , video redundancy coding (VRC) and I-MB refresh) to address the tradeoff between coding efficiency and error robustness [1]. Although decoders may suffer from quality degradation to a certain degree caused by error/loss in the network, they do not require error/loss-free transmission. On the other hand, non-real-time applications (e.g., email) usually can not tolerate any loss/error. Table 1.1 summarizes characteristics of different multimedia applications.

Table 1.1: Different multimedia applications

delay/jitter	real-time	non-real-time
loss/error	tolerant	non-tolerant
Examples	voice conversation	email,FTP

1.1.2   Wireless Networks

Wireless networks can be divided mainly into two types, i.e. the local-area network (LAN) and the wide-area network (WAN), based on the geographical cell size, the allocated frequency band, channel characteristics, main target applications, and economic concerns [3].

Wireless LANs primarily target high speed data applications within a relatively small geographical area, as a complement of their wired counterparts for situations where wiring is difficult or impractical. The architecture could be infrastructure-based or ad-hoc, whereby terminals communicate with each other without the mediation of a fixed base station. On the other hand, wireless WANs , such as cellular, Personal Communication Systems (PCS) , and mobile data radio systems, cover a large geographical area, but provide relatively low data-rate services compared to wireless LANs.

Due to differences mentioned above, wireless LANs and WANs have been developed based on different physical and link layer protocols. For example, in the medium access control (MAC) sub-layer, most commercial wireless LANs are based on packet-switching random access protocols, while most cellular WANs are based on circuit-switching FDMA/TDMA/CDMA protocols. Consequently, QoS issues are different in these two types of wireless networks.

For wireless LANs, how to support different multimedia applications, including real-time traffic with delay constraint and non-real-time traffic with loss/error constraint, by an efficient MAC protocol and packet scheduling algorithm, is the major concern. Moreover, for ad hoc wireless networks, in which an effective routing algorithm is one of the great challenges, QoS-aware routing is important [4].

For wireless WANs, the current circuit-switching technology can support real-time voice calls with a delay/delay jitter bound, as long as the circuit dedicated to the call can be established. Non-real-time applications have also been provided by low rate data services [5] ,[6] , when there are channels not occupied by real-time calls. The major concern to provide QoS to multimedia applications in such wireless networks is the seamless service upon user's movement.

Inefficient utilization of network resources is the major disadvantage of current circuit-switched networks. For example, the circuit is in use during the full lifetime of a real-time connection, even when no conversation is taking place. This has led to the deployment of the packet-switching technology in future wireless networks. Thus, some of the QoS issues of wireless LANs are also relevant in packet-switched wireless WANs.

In this book, we consider wireless WANs, especially cellular networks, and focuses on the radio resource management mechanisms for QoS provisioning.

1.2   Resource Management Issues in Cellular Wireless Systems

Fig. 1.1 provides a conceptual architecture of a cellular wireless communication network, which consists of a fixed network part and a wireless access system. The fixed network part, through mobile switching centers (MSC) , provides connections between radio access ports, often named as base stations (BS) , which in turn provide the wireless connections to the mobile station (MS) located in their coverage area (called cells) . BSs are distributed over the geographical area where communication services are covered. Continuous service coverage within this service area is achieved by handoff , which is the seamless transfer of a call from one BS to the other as the mobile unit crosses cell boundaries.

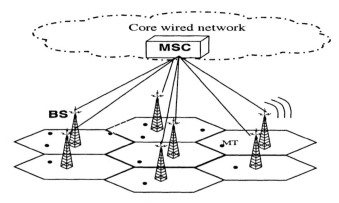

Figure 1.1: Illustration of the wireless network architecture.

1.2.1   Characteristics of Cellular Wireless Systems

The cellular systems have the following distinct characteristics.

- Cellular mobility and handoff
  A cellular network employs a central switching office, i.e. the MSC to interconnect small radio coverage areas into a larger system [7]. In order to maintain the service of a call, it is important to properly manage the resource along the path of MSs , especially when they move from one cell to another. Since each BS typically serves a limited number of users, additional customers cannot access the cellular system through that BS, when it has reached its capacity (i.e. all the radio channels are occupied). Therefore, an existing service may be disconnected during handoff due to the lack of the necessary resource in consecutive cells. To provide QoS guarantees to mobile users, a certain resource reservation scheme has to be implemented for potential handoff calls. This is the focus of the resource reservation module.

- Limited resource and preferential treatment
  With an increase in the number of users in a wireless communication system, the load on the frequency spectrum becomes heavier. Thus, it is important to manage the resource effectively. The resource under our consideration is either the number of channels (i.e. time slots or frequency narrow-bands) in 2G FDMA and TDMA systems or the maximum tolerable interference margin in 3G CDMA systems. When a system is congested, the preferential treatment should be given to higher priority users, to optimize the objective function of the system. This practice would serve the interests of the system operator as well as mobile users.

- Channel impairments
  Wireless channel between a MS and the BS is modeled as a time-varying communication path between the two stations. The point-to-point connection in this environment is affected by multipath, fast fading and other impairments, depending on propagation conditions. The "break-then-make" handoff process adopted by the FDMA/TDMA system suffers a great deal from channel impairments. In contrast, the soft handoff process in the CDMA system mitigates such a situation since

it allows an MS to connect to two or more BSs simultaneously during handoff.

1.2.2 Resource Management

Supporting the multiple QoS requirements of multimedia traffic is a major challenge in mobile wireless networks. The system-operator usually wants to achieve high system utilization so that more users can be accommodated by the system and more revenue can thus be obtained while the mobile user wants to receive better QoS. Consequently, a user would like the system to reserve enough resources for him/her to use, whenever required. In a wireless telephony system, squeezing more users into the system inevitably leads to congestion and heavier interference in the system. This results in poor QoS guarantees to users. Striking a proper balance between system utilization and user's QoS satisfaction is the focus of the schemes discussed in this book.

To meet the large bandwidth requirement of multimedia traffic, it is important to utilize the system resource efficiently and provide preferential treatment according to mobile user's traffic profile when the system is congested. The Radio Resource Management (RRM) module in the cellular network system is responsible for the management of air interface resources. RRM is needed to offer efficient system utilization and to guarantee a certain QoS to different users according to their traffic profiles. The call admission control (CAC) mechanism is one of the most important components of RRM that affects the resource utilization efficiency and QoS guarantees provided to users. The radio resource reservation estimation (RRE) mechanism helps CAC to decide how much resource is needed to be reserved in order to provide QoS guarantees to mobile users. The RRE module residing in each base station dynamically estimates the amount of resources to be reserved by referencing traffic conditions in neighboring cells periodically or upon the call request depending on the design of the system.

BIBLIOGRAPHY

[1] International Telecommunication Union, "Draft ITU-T H.263: Video coding for low bitrate communication," Draft International Standard, ITU-T, July 1997.

[2] ISO/IEC JTC1/SC29/WG11, "Overview of the MPEG-4 standard," ISO/IEC N3747, Oct. 2000.

[3] J. E. Padgett, C. G. Gunther, and T. Hattori, "Overview of wireless personal communications," *IEEE Communications Magazine*, 41(3):28–41, Jan 1995.

[4] S. Chakrabarti and A. Mishra, "QoS issues in ad hoc wireless networks," *IEEE Communications Magazine*, 39(2):142–148, Feb 2001.

[5] Y.-B. Lin, "Cellular digital packet data," *IEEE Potentials*, 16(3):11–13, Sept 1997.

[6] ETSI, "Digital cellular telecommunications system (phase 2+): General packet radio service (GPRS); overall description of the GPRS radio interface - stage 2 (GSM 03.64 v5.1.0)," TS 03 64 v5.1.0, ETSI, Nov 1997.

[7] L. Harte, R. Levine, and S. Prokup, *Cellular and PCS: The Big Picture*. McGraw-Hill Series, 1997.

Chapter 2

BACKGROUND

2.1 Evolution of Wireless Communication Networks

Wireless communications have experienced an enormous amount of growth during the last two decades. Here, we review the evolution of wireless communication systems and point out the improvements from one generation to the other.

2.1.1 The First Generation (1G) Systems

The first-generation (1G) wireless communication systems that used analog transmission for speech services were introduced in early 1980s. They included the Advanced Mobile Phone Service (AMPS) , the Total Access Communications System (TACS) and the Nordic Mobile Telephone (NMT) , etc. All of these 1G cellular systems used the frequency division multiple access (FDMA) method to achieve spectrum sharing among multiple users. To allow transmitting and receiving signals at the same time, the base station communicates with each mobile station with two separate sets of channels. The base station transmits along one set of radio channels, called forward channels, and receives along another set of channels, which are reverse channels from the mobile stations. A wireless system adopts the frequency reuse strategy (reuse pattern) to increase the number of radio channels. As cellular systems evolved, directional antennas are applied to sector a cell, to minimize the interference.

2.1.2 The Second Generation (2G) Systems

To meet the growing need for increasing the capacity of the cellular system, and to establish compatibility with the evolution of wired

networks towards digital systems, the second-generation (2G) wireless cellular systems based on digital transmission techniques were
introduced in late 1980s. Digitization allows the use of time division multiple access (TDMA) and code division multiple access
(CDMA) as alternative methods of radio resource sharing. Digital
cellular systems fall into three basic types of cellular technologies:
frequency division multiple access (FDMA), time division multiple
access (TDMA) and code division multiple access (CDMA). The
IS-136 system is sometimes referred to as Digital AMPS (DAMPS)
; it is the new generation of the TDMA system beyond the IS-54
dual-mode analog-digital system used in North America. IS-136 was
led by the Universal Wireless Communications Consortium (UWCC)
and Committee T1 (T1) sponsored by the Alliance for Telecommunications Industry Solutions, and was accredited by the American
National Standards Institute (ANSI). The Global System for Mobile Communications (GSM) developed in Europe is a mixed type
of TDMA/FDMA system. The development of GSM was led by the
GSM Association and the European Telecommunications Standards
Institute (ETSI) . CDMA systems differ from FDMA and TDMA
systems through the use of coded radio channels. IS-95 is an example of the CDMA technology, which was developed by the CDMA
Development Group (CDG) and Telecommunications Industry Association (TIA) . Personal Digital Cellular (PDC) is a 2G system as
well.

The 2G digital system has many advantages over the 1G system
in terms of capacity, quality, flexibility, security, and system complexity. With the demand for new innovative services in general,
and wide-band multimedia services in particular, the currently deployed 2G wireless systems have further evolved towards the third-
generation (3G) systems to offer more advanced service features.
Some systems that extend the existing 2G system are called 2.5G
systems. The main feature of 2.5G systems is the data packet service enhancement. They are developed to bridge the 2G and the
3G systems. One example of the 2.5G system is the General Packet
Radio Services (GPRS) that can provide higher data-rate packet-
switching services up to 115Kbps.

2.1.3   The Third Generation (3G) Systems

The 3G wireless systems have been standardized by the International Telecommunications Union (ITU) and the standard is known as International Mobile Telecommunications-2000 (IMT-2000) [1]. IMT-2000 is a unified technology standard that would provide a high quality worldwide roaming capability on a small terminal and support multimedia applications such as Internet browsing, e-commerce, e-mail, video conferencing , and access to information stored through wired networks. It will provide access, by means of one or more radio links, to a wide range of services supported by fixed telecommunication networks (e.g. PSTN/ISDN/IP) and to other services which are specific to mobile users.

It is envisioned that 3G IMT-2000 wireless networks will evolve from existing wireless systems through a phased approach. The first phase includes the circuit- and packet-switched networks and multimedia services supported by user bit rates up to approximately 2Mbps. Second phase is envisaged as augmenting first phase with additional service capabilities.

The first phase of IMT-2000 makes the best use of existing resources such as 2G mobile telecommunication standards IS-95 (CDMA) and GSM (TDMA) to provide high-speed wireless multimedia services. There are two kinds of approaches to this. One is the step-by-step asynchronous approach of W-CDMA evolving from GSM networks, led by the Third Generation Partnership Project (3GPP) in Europe [2]. With this approach, the 2G HSCSD (High Speed Circuit Switched Data) technique [3] can be used, where multiple channels are employed to support transmission of data switched by high-speed circuits. The maximum transmission rate is 57.6 Kbps in GSM/HSCSD specification. Furthermore, the General Packet Radio Services (GPRS) [4] is used to provide higher data rate packet switching services at up to 115Kbps, leading to Enhanced Data Rates for GSM Evolution (EDGE) [5], which can support services with a data rate up to 384 Kbps. This intermediate stage is called the 2.5 generation (2.5G). It will be finally upgraded to the 3G mobile communication system IMT-2000 [6]. The other is the step-by-step synchronous approach of CDMA-2000 evolving from current IS-95

networks, led by the 3GPP2 effort in North America. With this
approach, the maximum transmission speed of 2G is 64Kbps in IS-
95A/B specification while the maximum transmission speed of 2.5G
is up to 384 Kbps in the IS-2000 (1X) specification. The maximum
transmission speed of 3G is up to 2Mbps [7].

The key features of IMT-2000 include the following [7, 8, 13]:

- bit rates up to 2 Mbps (high bit rate transmission);

- a variable bit rate to offer bandwidth on demand (rate adaptivity);

- high quality requirements from 10% frame error rate to 10^{-6}
  bit error rate (a wide range of QoS requirement);

- high spectrum efficiency (via radio resource management and
  other mechanisms);

- multiplexing of services with different quality requirements
  (multimedia support);

- support of packet-based transmission.

IMT-2000 provides a framework for worldwide wireless access by
integrating a diverse system consisting of both terrestrial and satellite networks. It also exploits the potential synergy between digital mobile telecommunication technologies and fixed wireless access
(FWA) Systems [9].

2.2   Challenges in Wireless Multimedia Networks

Providing multimedia services in wireless communication systems
will become a reality in the very near future. To achieve this goal,
many technical challenges are yet to be met as discussed in this
section.

2.2.1   Radio Aspect

Higher rate services with high quality to be provided in the future
wireless multimedia networks will demand more spectrum-efficient
radio technologies [11].

- Radio Interface Access
  The medium access technology to the common radio spectrum should maximize the system capacity while minimizing the interference among users. Moreover, it should enable the air interface to be able to cope with variable asymmetric data rates of multimedia services with different bandwidth on demand.

- Radio Transmission Technology
  Radio transmission technologies, including modulation and spreading, channel equalization and coding, smart/adaptive antenna systems, etc., should be flexible and spectrum efficient, and provide high quality service with a reasonable complexity.

- Radio Resource Management
  The rapid increase in the size of the wireless mobile community and its demands for high-speed multimedia communications stand in clear contrast to the rather limited spectrum resource. Efficient spectrum (or radio resource) management [25] is of great importance for assigning channels and the transmitter power to the radio access port and the terminal under the interference constraint.

2.2.2   Network Aspect

Technical issues in the network aspect address the functional architecture, signaling and protocols of 3G wireless multimedia networks [26]. Some key issues are as follows.

- Mobility Management
  Mobility management is one of the most important issues in wireless networks in order to support user mobility [27]. There are two aspects of mobility management for wireless WANs, i.e. handoff management and roaming management. Handoff management enables a mobile node to seamlessly obtain transmission services after handoff to a new cell. Roaming management, including registration and location update, enables service delivery to users who move to the coverage area of another network. Handoff management includes issues such

as handoff detection, radio link transfer, and channel assignment. To meet the requirements of increased system capacity, high signal quality, and power saving in mobile terminals, a reduced cell size will be deployed in the future wireless networks. This leads to an increased number of handoff, which makes handoff management very important. Furthermore, with integrated voice, data and video services, mobility management in the future mobile networks has to also address issues such as routing.

- Signaling and Protocol
  Signaling and protocols that are necessary for service access, call/connetion establishment, maintenance and release, etc., will require enhancement to support multimedia applications in the presence of user mobility and change of user profiles.

- Security and Privacy
  Security is becoming increasingly important in wireless and mobile telecommunication services to protect end-user's privacy and to protect network operators from significant loss of revenues due to fraudulent access to networks and services.

2.2.3 Other Aspects

Other challenges faced by 3G wireless multimedia networks include the development of easy-to-use multimedia handset terminals with simple attractive user interfaces, convenient operations, lower energy consumption, provision of various applications and contents, etc. To meet these challenges, related technology facilitators are envisioned as but not limited to the following:

- Battery technologies

- Integrated Circuit (IC) technologies

- Digital signal processing and compression

- Multimedia content access and retrieval

As a summary, provision of multimedia services in the future wireless networks has prompted a wide range of technology developments. These technologies should not only enable 3G wireless multimedia networks, but also support QoS, which is a key factor to the success of 3G wireless multimedia services.

2.3 Challenges in Radio Resource Management

The radio transmission technology and the computer network technology are two core components in a cellular communication system. From the perspective of call connection, a MS will experience some discontinuity due to handoff, when it moves from one cell to another. Unlike the wired-line telephone system, a cellular system has to consider the impairment of air-interface and the change of connection due to the mobility of MS.

With growing demand for more and more QoS features and multimedia support in the next generation cellular system, the radio resource management (RRM) has become crucial. RRM is the process of developing decisions and taking actions to optimize the system utilization. Technical challenges in RRM are addressed below.

2.3.1 Handoff and Mobility Issues

In cellular wireless networks, the geographical territory is divided into several regions and cells to enable the frequency reuse. This also ensures a higher system capacity. Thus, each BS can provide services only to a limited number of MS. As an MS is not attached to a fixed infrastructure, it can move from one cell to the other, and handoff occurs. The cellular system needs to track and locate MS accordingly.

Moreover, the cellular system needs to allocate or reserve resource in advance in order to maintain connection or certain QoS requirement for that connection. In the handoff process, such tasks are performed by resource management, which decides how much resource are needed to be allocated or reserved for MS in order to achieve the QoS objectives. (Here, the resource in the TDMA/FDMA system is referred to as the number of available channels. In the CDMA system, the resource corresponds to the available interference margin that a system can tolerate.)

2.3.2   Channel Assignment and Reservation

There are many ways of allocating the resources, in response to a new call arrival or a handoff attempt. Here the goal is to maximize system efficiency while meeting the QoS requirement of users. Some channel assignment techniques allow channel sharing among several cells. Such schemes, known as Dynamic Channel Assignment (DCA) [14, 15] schemes, have a common resource sharing pool that can be allocated to users upon request. In DCA, channels are not permanently assigned to cells. These schemes can borrow channels from neighboring cells if necessary. However, once a channel is borrowed from a neighboring cell, all other cells that are within the co-channel reuse distance are prohibited from using the channel. Therefore, its performance could be worse in the heavy traffic scenario. The other channel assignment schemes allocate a fixed amount of resource to each cell. They are referred to as static channel assignment (SCA) schemes. Most practical wireless systems belong to the SCA category since the complexity of DCA is much higher than that of SCA. Our work falls into the SCA category due to its simplicity. Our proposed schemes give preferential treatment to priority classes by using appropriate resource reservation schemes.

2.3.3   Preferential Treatment

Due to a limited amount of resources, a system may not be able to support services that meet requirements of all users at the same time. Users who want to pay more or are considered more important can be given preferential treatment by assigning a higher priority to receive a certain QoS guarantee. One difficulty here is how to choose the proper weighting among priorities. This is rather a subjective issue.

Two common strategies are employed to provide preferential treatment to handoff calls as compared to the new calls. They are the handoff queue (HQ) [16] and the guard channel(GC) [17, 18] schemes. HQ-based methods follow the following principle. When the resource becomes available, one of the calls in the handoff queue is served. If there is no available resource, call requests are queued until the resource becomes available again. The HQ scheme needs a large buffer and a sophisticated scheduling mechanism to provide required QoS to real-time multimedia traffic so as to ensure that queued data do not expire. The basic idea of GC-based admission control schemes is to reserve resources known as guard channels a priori in each cell and to give preferential treatment to high-priority and/or handoff calls. As a result, the scheme offers already admitted mobile users a better connectivity than users requesting a new call, especially during heavy traffic. In such a system, call requests of a lower priority are rejected if the available resource is less than a certain threshold.

Detailed implementations of the GC scheme may differ in the number of guard channels reserved by a base station. Hong and Rappaport [17] used a fixed GC scheme to treat new calls and hand-off calls differently by reserving the same amount of resource for the handoff calls in the entire period of the simulation cycle. In this work, only one traffic class was considered. In [19], Rapport and Purzynski extended the work to multiple service classes. They analyzed the performance based on a proposed mathematical model with the assumption of stationary traffic. Epstein and Schwartz [20] considered a mixed traffic with calls of narrow and wide-band. We discuss a scheme in Chapter 6.1 that extends the single threshold in the fixed GC scheme to multiple thresholds to deal with multimedia traffic with different priorities. All schemes discussed above are static since they cannot adapt to the quick variation of traffic patterns.

Many dynamic GC schemes have also been discussed in the literature to improve system efficiency while providing the QoS guarantees to high-priority calls. These dynamic schemes adaptively reserve the amount of resources needed for high-priority calls and, therefore, accept more lower priority calls as compared to the fixed scheme. Naghshineh and Schwartz [21] proposed an analytical model to estimate the resource requirements for handoff calls. In their model, all connection requests have an identical traffic profile and the traffic is under stationary conditions. Ramanathan *et al.* [22] proposed a dynamic resource allocation scheme by estimating the maximal expected resources needed for handoff calls. In [23], Acampora *et al.* applied a linear weighting scheme (LWS) as a part of their admission control algorithm that uses the average number of ongoing calls in all neighboring cells within the region of awareness to determine the admission. Sutivong and Peha [24] adopted a hybrid scheme based on the weighted sum of ongoing calls in the originating cell and in other cells to determine the call admission policy.

2.4   QoS in Wireless Multimedia Networks

The experience with multimedia applications over wired networks has demonstrated the need for QoS. However, providing multimedia services with QoS guarantee in wireless environment presents more challenges.

First, the link bandwidth is limited in a wireless environment. Although IMT-2000 is envisaged to be able to support user bit rates up to 2 Mbps [26], the bandwidth is still limited for emerging and expanding multimedia services compared to broadband wired networks such as ATM, especially in the presence of some bandwidth intensive multimedia applications (e.g. video and audio). Thus, the bandwidth of a wireless network is still a bottleneck.

Second, the wireless channel is highly variable due to many reasons. The characteristics of a radio channel changes dramatically and rapidly due to phenomena such as multi-path scattering from nearby objects, shadowing from dominant objects, signal attenuation and fading, even if a wireless host is not mobile. Consequently, the received signal power and the delivery delay of signals between the sender and the receiver vary with the location and time, which makes QoS guarantee more difficult than that in wired networks.

Third, QoS guarantee of wireless multimedia is further complicated by user's mobility. Achieving location transparency (i.e. QoS guarantee at any time wherever a mobile host moves) is a challenging problem. In a wired network, it is sufficient to establish and guarantee that the end-to-end path has adequate resources to deliver an information flow between end users with negotiated QoS. The situation in a wireless mobile network is much more volatile, because a mobile node's end-to-end path is likely to change as it moves through the network, and the wireless link is subject to wide fluctuations in its performance and reliability. Resource reservation is thus complicated by the need to consider resource availability not only on the initial end-to-end path but also on other paths likely to be used as the node position changes.

Therefore, supporting QoS in wireless networks is extremely complex due to the limited bandwidth resources, highly variable environment, and user's mobility. To address this problem, we consider QoS in wireless networks at two abstraction levels: *connection-level QoS* and *application-level QoS*. The functionality and involved technologies of each level QoS are detailed in the following sections.

2.4.1   Connection-level QoS

The basic level of QoS over a wireless link is the connection-level QoS, which is related to connection establishment and management. The connection-level QoS is very important due to the distinct characteristics of wireless networks, especially in the presence of user mobility. It measures the service connectivity and continuity of a wireless network, and provides the basis of application-level QoS.

Connection-level QoS is often measured by two parameters, i.e. *the new call blocking probability*, which measures service connectivity, and *the handoff dropping probability*, which measures service continuity during handoff. The new call blocking probability is the probability of a new call request being rejected due to unavailability of resources to support this call. The handoff dropping probability is the probability of an on-going call being forced-terminated before its completion, when a mobile user moves to a new cell during the call's lifetime and the new cell does not have enough resources to support it.

For a mobile user, dropping an on-going call is generally more unbearable than blocking a new call request. Therefore, minimizing the call dropping probability is a major objective in the wireless system design. On the other hand, the goal of a network service provider is to maximize the revenue by improving network resource utilization, which is however tied with minimizing the call blocking probability.

For connection-level QoS, mobility management, especially hand-off management, is an important issue. How to switch connections efficiently upon handoff is a major concern due to the importance of reducing handoff droppings. The solution involves radio resource allocation, call admission and resource reservation for handoff calls.

2.4.2   Application-level QoS

Connection-level QoS is necessary but usually not sufficient. Application-level QoS is related to perceived quality by end users, whose service requests are connected and continued through connection-level QoS support. There are a set of parameters describing application-level QoS, such as delay/delay jitter, error/loss performance, throughput, etc.

Application-level QoS is particularly critical in packet switched networks, which take advantage of a higher degree of multiplexing among services. As a result, packets for a certain service flow may experience varying delay, delay jitter and loss. Application-level QoS is also referred to packet-level QoS here. Efficient packet access protocols and packet scheduling schemes are key components of solutions to these QoS issues. In order to transfer bursty real-time and non-real-time data, flexible allocation of radio resources for packet-type services with unpredictable bit rates, taking into account fairness between asymmetric data transmission, is an essential requirement for 3G networks. Channel coding and power control are also essential to the error performance. The challenge is to achieve the required flexibility without an overwhelming complexity in the network and terminals.

The two QoS levels are related to each other in the sense of resource allocation, which considers allocation among connection flows and among new and handoff calls, through proper call admission control and resource reservation schemes. On the other hand, from the viewpoint of end users, QoS should be provided on an end-to-end basis. Thus, in a wireless multimedia network with infrastructure as illustrated in Fig. 1.1, QoS over wireless access networks needs interaction and mapping with that over the core wired networks.

2.4.3   Related Work on QoS in Wireless Networks

Different techniques have been proposed to support QoS for multi-media applications at different layers of wireless networks. At the application layer, some multimedia coding systems, such as the H.263L video codec [28], are targeting transmission over wireless links. At the network layer, techniques of mobility management and seamless connectivity are used. Routing mechanisms are also extended to be QoS aware and able to handle mobility [29]. At the data link layer, medium access control is modified so that reservations and QoS guarantees can be supported. At the physical layer, modulation and power control schemes are also designed to be QoS aware [30].

A wireless network is considered a highly variable environment. To maintain a specified QoS level, when the wireless link fluctuates or degrades, a wireless system has to adapt to varying conditions. Adaptability of wireless multimedia networks can also be made at different layers. At the application layer, most recent real-time applications are made to be able to adapt to changing networking conditions. At the network layer, routing methods should adapt to mobility. At the data link layer, error correction mechanisms can be adaptive to protect against higher and varying error rates of wireless links. Adaptability at the physical layer is possible by choosing appropriate channel and power control techniques.

Previous Work on Connection-level QoS Provisioning

The fundamental difference between wireless radio channel and the wired link, and user's mobility make the connection-level QoS more important in wireless networks. To leverage the two connection-level QoS measurements, different CAC and resource reservation schemes have been proposed for wireless/mobile networks.

One of the first bandwidth reservation schemes, known as the *guard channel scheme* for handoff was introduced in mid 1980's by

Hong and Rapport [17]. In this scheme, a set of channels are permanently reserved exclusively for handoff calls to keep the handoff dropping probability lower than the new call blocking probability. It has been shown that this reservation scheme was the optimal stationary scheme to minimize a linear objective function of the two probabilities under certain assumptions [31]. However, it cannot provide any guarantee on the connection-level QoS. Ramjee *et al.* [31] also investigated the problem of minimizing the new call blocking probability (and thus maximizing the bandwidth utilization) subject to a predefined target handoff dropping probability and found the optimal stationary fractional guard channel scheme. However, the static reservation in the stationary CAC schemes is not efficient for varying traffic conditions found in wireless networks.

Several distributed CAC schemes have been proposed [32, 33, 34] to dynamically calculate the required bandwidth in order to maintain a limited cell overload probability. These approaches use statistical modelling analysis to estimate the probability by making assumptions about some parameters, such as the call holding time, the cell residence time, the handoff rate, etc. However, the performance depends on the conformance of the real-system to the model used and the accuracy of those assumptions made for the model parameters. Generally, a real network system cannot be approximated by a simple model without making some unrealistic assumptions. Thus, more elaborate models are usually needed to make better estimation. However, the more elaborate the model is, the more complex it is to analyze and the more sensitive it is to the accuracy of assumptions.

The concept of *shadow cluster* for resource reservation and admission control was introduced by Levine *et al.* [35] to reduce the call dropping probability by predictive resource allocation. In this scheme, a shadow cluster is a set of cells around an active mobile. However, how to determine the shadow cluster is not defined clearly. This scheme requires base stations in neighboring cells to predict future resource demands according to the information of the bandwidth requirement, the position, the movement pattern and the time of mobile users. As a result, it is computationally too expensive to be practical.

More recently, a few measurement-based dynamic CAC schemes

have been proposed [36, 37]. These approaches rely on the measurement of current traffic conditions within a number of cells. Therefore, either information exchange among neighboring cells [36] or accurate location-tracking for all users [37] are required. Moreover, these schemes are based on traditional mobile networks with only one type of traffic and cannot effectively handle a variety of connection bandwidth, traffic loads, and user's mobilities.

Previous Work on Application-level QoS Provisioning

There has also been research effort on application-level QoS. Most of existing work focuses on packet-level QoS in Wireless Local Area Networks (WLANS), which is emerging as an attractive alternative or complementary to wired LANs. The time-slot allocation based wireless medium access control (MAC) protocol is a popular method to support packet-level QoS guarantees in WLANs. Existing packet scheduling algorithms for wired networks have been investigated to be used in wireless LANs for different traffic types with different delay and/or loss requirements in [38], [39], [40]. Channel errors are also addressed by using hybrid FEC/ARQ schemes [40]. A WLAN MAC standard, IEEE 802.11, provides both real-time and non-real-time communication services in a WLAN using polling and CSMA/CA [41], but does not specify other issues such as error control.

The above-mentioned schemes do not address the handoff and mobility management issues that are inevitable in the cellular environment. In other words, they consider the packet-level QoS issue only without addressing the connection-level QoS issue at all. Therefore, they are not sufficient for 3G wireless multimedia networks, which is the major target of our research.

Previous Work on Both Connection-level and Application-level QoS Provisioning

Several researchers have addressed both connection-level and application-level QoS for wireless networks. Zhuang et al. [42] described an adaptive QoS handoff priority scheme to reduce the handoff dropping. The connection-level and packet-level performance in terms of delay was studied analytically for a homogeneous network supporting two kinds of service types, namely wide-band and narrow-band services. However, it deals with the two types of QoS issues separately. Thus, there is no guarantee or optimization on the QoS performance of either level.

Oliveira *et al.* [43] proposed an admission control scheme for multimedia applications by considering both level QoS. For application-level QoS, it assumed two types of traffic, i.e. the real-time traffic and the non-real-time traffic. For connection-level QoS, this scheme dynamically reserves the bandwidth in cells surrounding a cell where the connection is established to provide QoS guarantees to real-time traffic in high-speed multimedia wireless networks. When a user moves to a new cell, the reserved bandwidth is utilized. The bandwidth reservation process is repeated in new neighboring cells, and the reserved bandwidth in cells, which are no longer in the neighborhood of the current cell, is released. However, the bandwidth reservation in all neighboring cells is apparently a waste of resource as the mobile user hands off to only one of them.

Talukdar *et al.* [44] proposed an admission control scheme based on a new three-class service model for integrated services packet networks with mobile hosts. It extends the InteServ architecture [45] proposed for QoS provisioning in wired IP networks to mobile hosts. This scheme considered the delay bound caused by packet scheduling algorithms as the packet-level QoS metrics. There is a limitation in the scheme since it assumes that each new mobile user should provide its accurate mobility specification, which consists of the set of cells the mobile host is expected to visit during the lifetime of the flow. This limits the flexibility gained from mobility.

The research work in this monograph addresses both the connection-level QoS and the packet-level QoS for future packet-switched wireless multimedia networks. For packet-level QoS, a service model with three service classes, *i.e.* real-time constant-bit-rate, real-time variable-bit-rate and non-real-time unspecified-bit-rate, is proposed. The call admission control schemes for real-time CBR and VBR services are developed to jointly optimize the two level QoS.

BIBLIOGRAPHY

[1] E. Dahlman, B. Gudmundson, M. Nilsson, and J. Skold, "UMTS/IMT-2000 based on wideband CDMA," *IEEE Communications Magazine*, 36(9):70–80, Sept 1998.

[2] T. Ojanpera and R. Prasad, "An overview of third-generation wireless personal communications: A European perspective." *IEEE Personal Communications*, 5(6):59–65, Dec 1998.

[3] ETSI, "Digital cellular telecommunications system (phase 2+): High speed circuit switched data (HSCSD) - stage 2 (GSM 03.34)," Technical report, ETSI. TS 101 038 V5.0.1, Apr 1997.

[4] ETSI, "Digital cellular telecommunications system (phase 2+): General packet radio service (GPRS); overall description of the GPRS radio interface - stage 2 (GSM 03.64 v5.1.0)," TS 03 64 v5.1.0, ETSI, Nov 1997.

[5] K. Zangi, A. Furuskar, and M. Hook, "Edge: Enhanced data rates for global evolution of GSM and IS-136," In *Proc. of Multi Dimensional Mobile Communications (MDMC'98)*, 1998.

[6] N. R. Prasad, "GSM evolution towards third generation UMTS/IMT2000," In *Proc. of IEEE International Conference on Personal Wireless Communication*, pp. 50–54, Feb. 1999.

[7] Y. S. Rao and A. Kripalani, "CDMA2000 mobile radio access for IMT2000," In *Proc. of IEEE International Conference on Personal Wireless Communication*, pp. 6–15, Feb 1999.

[8] T. Ojanpera and R. Prasad, "An overview of third-generation wireless personal communications: a European perspective," *IEEE Personal Communication Magazine*, pp. 59-65, Dec. 1998.

[9] ITU, "International mobile telecommunications-2000," *http://www.itu.org*, 2000.

[10] 3GPP2, "The third generation partnership project 2 (3GPP2),"
     http://www.3gpp2.org, 2000.

[11] R. D. Carsello, R. Meidan, S. Allpress, F. O'Brien, J. A. Tarallo,
     N. Ziesse, A. Arunachalam, J. M. Costa, E. Berruto, R. C.
     Kirby, A. Maclatchy, F. Watanabe, and H. Xia, "IMT-2000
     standards: Radio aspects," *IEEE Personal Communications*,
     4(4):30–40, Aug 1997.

[12] K. Richardson, "UMTS overview," *Electronics &Communica-
     tion Engineering Journal*, vol. 12, pp. 93–100, June 2000.

[13] H. Aghvami and B. Jafarian, "A vision of UMTS/IMT-2000
     evolution," *Electronics &Communication Engineering Journal*,
     vol. 12, pp. 148–152, June 2000.

[14] Z. Zheng and W. Lam, "Analytical methods to calculate perfor-
     mance of handoff prioritisation in dynamic channel assignment,"
     Electronics Letters, vol. 37, pp. 978–979, July 2001.

[15] S. Boumerdassi, "An efficient reservation-based dynamic chan-
     nel assignment strategy," *First International Conference on 3G
     Mobile Communication Technologies*, pp. 352–355, 2000.

[16] P. O. Gaasvik, M. Cornefjord, and V. Svensson, "Different
     methods of giving priority to handoff traffic in a mobile tele-
     phone system with directed retry," *41st IEEE Vehicular Tech-
     nology Conference, Gateway to the Future Technology in Mo-
     tion*, pp. 549–553, 1991.

[17] D. Hong and S. S. Rapport, "Traffic model and performance
     analysis for cellular mobile radiotelephone systems with prior-
     itized and nonprioritized handoff procedures," *IEEE Transac-
     tions on Vehicle Technology*, vol. 35, pp. 77–92, Aug. 1986.

[18] T. Kwon, Y. Choi, C. Bisdikian, and M. Naghshineh, "Call
     admission control for adaptive multimedia in wireless/mobile
     networks," in *IEEE Wireless Communications and Networking
     Conference*, pp. 540–544, 1999.

[19] S. S. Rapport and C. Purzynski, "Prioritized resource assignment for mobile cellular communication systems with mixed services and platform types," *IEEE Transactions On Vehicular Technology*, vol. 45, pp. 443–458, Aug. 1996.

[20] B. Epstein and M. Schwartz, "Reservation strategies for multimedia traffic in a wireless environment," in *Vehicular Technology Conference, 1995 IEEE 45th*, (Chicago, IL), pp. 165–169, July 1995.

[21] M. Naghshineh and M. Schwartz, "Distributed call admission control in mobile/wireless networks," *IEEE Journal Selected Areas Communications*, vol. 14, pp. 711–717, May 1996.

[22] P. Ramanathan, K. M. Sivalingam, P. Agrawal, and S. Kishore, "Dynamic resource allocation schemes during handoff for mobile multimedia wireless networks," *IEEE Journal on Selected Areas in Communications*, vol. 17, pp. 1270–1283, July 1999.

[23] A. S. Acampora and M. Naghshineh, "Control and quality of service provisioning in high-speed micro-cellular networks," *IEEE Personal Communications*, No. 2nd Quarter, pp. 36–43, 1994.

[24] A. Sutivong and J. M. Peha, "Novel heuristics for call admission control in cellular systems," *1997 IEEE 6th International Conference on Universal Personal Communications Record*, vol. 1, pp. 129–133, 1997.

[25] J. Zander, "Radio resource management in future wireless networks: requirements and limitations," *IEEE Communications Magazine*, 35(8):30–36, Aug 1997.

[26] R. Pandya, D. Grillo, E. Lycksell, P. Mieybegue, H. Okinaka, and M. Yabusaki, "IMT-2000 standards: network aspects," *IEEE Personal Communications*, 4(4):20–29, Aug 1997.

[27] B. Jabbari, G. Colombo, A. Nakajima, and J. Kulkarni, "Network issues for wireless communications," *IEEE Communications Magazine*, 33(1):88–98, Jan 1995.

[28] International Telecommunication Union, "Draft ITU-T H.263: Video coding for low bitrate communication," Draft International Standard, ITU-T, July 1997.

[29] S. Chen and K. Nahrstedt, Distributed quality-of-service routing in ad-hoc networks. *IEEE Journal on Selected Areas in Communication*, 17(8):1488–1505, Aug 1999.

[30] L. Qiu, P. Xia, and J. Zhu. "Study on wideband CDMA modulation, power control and wireless access for CDMA multimedia systems," In *Proc. of IEEE VTC '99*, vol. 5, pp. 2944–2948, Sept 1999.

[31] R. Ramjee, D. Towsley, and R. Nagarajan, "On optimal call admission control in cellular networks," *Wireless Networks*, 3(1):29–41, May 1997.

[32] M. Naghshineh and S. Schwartz, "Distributed call admission control in mobile/wireless networks," *IEEE Journal on Selected Areas in Communication*, 14(4):711–717, May 1996.

[33] S. Wu, K. Y. M. Wong, and B. Li, "A new distributed and dynamic call admission policy for mobile wireless networks with QoS guarantee," In *Proc. of the Ninth IEEE International Symposium on Personal, Indoor and Mobile Radio Communications*, vol. 1, pp. 260–264, Sept. 1998.

[34] T. Kwon and Y. Choi, "Call admission control for mobile cellular network based on macroscopic modeling of vehicular traffic," In *Proc. of IEEE VTC*, pages 1940–1944, May 1998.

[35] D. Levine, I. Akyildiz, and M. Naghshineh, "A resource estimation and call admission algorithm for wireless multimedia networks using the shadow cluster concept," *IEEE/ACM Trans. on Networking*, 5(1):1–12, Feb. 1997.

[36] J. M. Peha and A. Sutivong, "Admission control algorithms for cellular systems," *Wireless Networks*, 7(2):117–125, Apr 2001.

[37] W. S. Soh and H. S. Kim, "Dynamic guard bandwidth scheme for wireless broadband networks," In *Proc. of IEEE INFOCOM 2001*, vol. 1, pp. 572–581, Apr 2001.

[38] J. M. Capone and I. Stavrakakis, "Delivering diverse delay/dropping QoS requirements in a TDMA environment," In *Proc. of ACM MobiCom'97*, volume 1, pages 110–119, Sept. 1997.

[39] S. Choi and K. G. Shin, "A cellular wireless local area network with QoS guarantees for heterogeneous traffic," In *Proc. of IEEE INFOCOM'97*, volume 3, pages 1030–1037, Apr. 1997.

[40] N. R. Figueira and J. Pasquale, "Providing quality of service for wireless links: wireless/wired networks," *IEEE Personal Communications*, 6(5):42–51, Oct. 1998.

[41] B. P. Crow, I. Widjaja, J. G. Kim, and P. T. Sakai, "IEEE 802.11 wireless local area networks," *IEEE Communications Magazine*, 35(9):116–126, Sept 1997.

[42] W. Zhuang, B. Bensaou, and K. C. Chua, "Adaptive quality of service handoff priority scheme for mobile multimedia networks," *IEEE Trans. on Vehicular Technology*, 49(2):494–505, Mar 2000.

[43] C. Oliveira, J. B. Kim, and T. Suda, "An adaptive bandwidth reservation scheme for high-speed multimedia wireless networks," *IEEE Journal on Selected Areas in Communication*, 16(6):858–874, Aug. 1998.

[44] A. K. Talukdar, B. R. Badrinath, and A. Acharya, "Integrated services packet networks with mobile hosts: Architecture and performance," *Wireless Networks*, 5(2):111–124, Mar. 1999.

[45] R. Braden, D. Clark, and S. Shenker, "Integrated service in the internet architecture: an overview," IETF RFC1633, 1994.

Chapter 3

ANALYSIS OF CONNECTION-LEVEL QOS

In this chapter, mathematical models of wireless multimedia networks are studied to provide a foundation of connection-level QoS analysis. We start from traditional wireless communication networks with only one type of application (voice). In this case, all calls require the same bandwidth resource. This kind of wireless network has been widely studied with basic queuing theory. Two different schemes, i.e. non-priority handoff and priority handoff, are modeled to analyze connection-level QoS. Then, we extend the modeling and analysis to wireless multimedia networks with different types of applications requiring different amount of bandwidth resources, for both non-priority and priority handoff schemes. Finally, a promising feature of emerging multimedia applications called rate-adaptability is considered, and rate-adaptive models for non-priority and priority handoff are analyzed.

3.1 Homogeneous Bandwidth Applications

Traditional wireless communication networks only have the voice application where each call demands the same amount of bandwidth, denoted as one channel. Considering a single cell with a fixed amount of bandwidth capacity (e.g. C channels), we make the following assumptions.

- The new call request to this cell is a Poisson process with rate λ_n.

- The handoff request to this cell is a Poisson process with rate λ_h.

- The time for each call staying in this cell is exponentially distributed with mean cell residence time $\frac{1}{\eta}$.

- Each call requires one channel bandwidth.

- The arrivals of new call and handoff call requests are independent of each other.

Based on the above assumptions, a cell with homogeneous bandwidth applications can be modeled by a M/M/C/C queuing system. The first M denotes that the inter-arrival time has exponential distribution, which is equivalent to Poisson arrival, and the second M denotes the service time has also an exponential distribution. The first C indicates that the queuing system has C servers, which are the C channels in the cell. The second C indicates that the system has a capacity of C, which means there are at most C customers in the system at any time. In this wireless network scenario, if a call request arrives when there are already C calls in the cell, i.e., all C channels are busy, then it will be blocked. The state of the queuing system x is defined as the number of occupied channels in the cell, which is the same as the number of calls in the cell. Then, the state space is $\Lambda = \{x, |0 \leq x \leq C\}$.

Basically, the M/M/C/C queuing model is a continuous-time Markov chain with C discrete states. Based on the assumption that each call occupies one channel, the transition of states only occurs between neighboring states. This kind of special Markov chain is also called the *birth-death process* , defined as a continuous-time Markov chain with transition probabilities given by

$$p(i, i+1) = a(i) \tag{3.1}$$
$$p(i, i-1) = b(i) \tag{3.2}$$
$$p(i, i) = 1 - (a(i) + b(i)) \tag{3.3}$$
$$p(i, k) = 0, \text{otherwise}, \tag{3.4}$$

where $p(i, k)$ is the transition probability from state i to state k. Birth occurs when there is one arrival that results in the transition from the current state to the next state. Death occurs when there is a departure that results in the transition from the current state to the previous state. Thus, transition rates from the current state x to the next state $x+1$ and to the previous state $x-1$ are also called the birth rate $\lambda(x)$ and the death rate $\mu(x)$, respectively.

3.1.1 Non-priority Handoff Scheme

The non-priority handoff scheme treats new call and handoff call requests the same by admitting a request (either from a new call or from a handoff call) as long as there is one unused channel in the current cell. Fig. 3.1 illustrates the state transition rate diagram of this scheme.

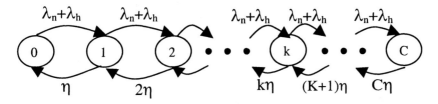

Figure 3.1: Illustration of the state transition diagram for the non-priority handoff scheme.

For the non-priority handoff scheme, the birth rate of state x can be written as

$$\lambda(x) = \begin{cases} \lambda_n + \lambda_h, & 0 \leq x < C, \\ 0, & \text{otherwise.} \end{cases} \qquad (3.5)$$

The death rate of state x is equal to

$$\mu(x) = \begin{cases} x\eta, & 0 < x \leq C, \\ 0, & \text{otherwise.} \end{cases} \qquad (3.6)$$

The steady state distribution, which gives the limiting probability $P(k)$ of the system staying in state k can be obtained by solving the following series of equations.

$$\lambda(k-1)P(k-1)+\mu(k+1)P(k+1) = (\lambda(k)+\mu(k))P(k), \qquad 0 \leq k \leq C \qquad (3.7)$$

and

$$\sum_{k=0}^{C} P(k) = 1 \tag{3.8}$$

Equation (3.7) is the equilibrium difference equation for each state, which comes from the fact that the input flow rate must equal to the output flow rate for a given state in equilibrium. Equation (3.8) is the constraint of the probability distribution function.

Then, the steady state distribution is found to be [1]

$$P(k) = \begin{cases} \frac{(\frac{\lambda_n + \lambda_h}{\eta})^k}{k!} P(0), & 0 < k \leq C \\ 0 & \text{otherwise,} \end{cases} \tag{3.9}$$

where

$$P(0) = \left(\sum_{i=0}^{C} \frac{(\frac{\lambda_n + \lambda_h}{\eta})^i}{i!} \right)^{-1}. \tag{3.10}$$

Given the steady state distribution, the new call blocking probability P_B and the handoff dropping probability P_D can be found as the probability that the system is in the state with all C channels occupied, i.e.

$$P_B = P_D = P(C). \tag{3.11}$$

It is obvious that the new call blocking probability and the handoff dropping probability are the same for the non-priority handoff scheme.

3.1.2   Priority Handoff Scheme

Unlike the non-priority handoff scheme, the priority handoff scheme gives a higher priority to the handoff call request than the new call request. Here, we consider the prevalent GC scheme, which reserves a portion of the channel resource to handoff calls only. More specifically, a new call request is admitted only when there are less than T channels occupied, where T is a threshold between 0 and C. On the other hand, a handoff request is rejected only if all C channels are occupied. As a result, there are $C - T$ channels, named *guard channels*, can be used by handoff calls only. With the same assumptions made in the non-priority handoff scheme, a cell deploying the

GC scheme can be modeled as a similar M/M/C/C queuing system
with a threshold state T as illustrated in Fig. 3.2

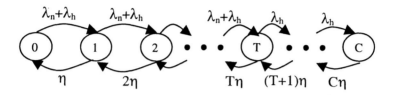

Figure 3.2: Illustration of the state transition diagram for the priority
handoff (GC) scheme.

For the GC scheme, the birth rate of state x becomes

$$\lambda(x) = \begin{cases} \lambda_n + \lambda_h, & 0 \le x < T, \\ \lambda_h, & T \le x < C, \\ 0, & \text{otherwise.} \end{cases} \tag{3.12}$$

The death rate of state x is

$$\mu(x) = \begin{cases} x\eta, & 0 < x \le C, \\ 0, & \text{otherwise.} \end{cases} \tag{3.13}$$

By using the same equilibrium method as used in the non-priority
handoff scheme, the steady state distribution can be derived and
given below. [1]

$$P(k) = \begin{cases} \dfrac{(\frac{\lambda_n+\lambda_h}{\eta})^k}{k!}P(0), & 0 \le k < T, \\ \dfrac{(\frac{\lambda_n+\lambda_h}{\eta})^T(\frac{\lambda_h}{\eta})^{k-T}}{k!}P(0), & T \le k \le C, \\ 0, & \text{otherwise,} \end{cases} \tag{3.14}$$

where

$$P(0) = \left(\sum_{i=0}^{T} \frac{(\frac{\lambda_n+\lambda_h}{\eta})^i}{i!} + \sum_{i=T+1}^{C} \frac{(\frac{\lambda_n+\lambda_h}{\eta})^T(\frac{\lambda_h}{\eta})^{i-T}}{i!} \right)^{-1}. \tag{3.15}$$

Given the steady state distribution, the new call blocking probability P_B is the probability that the system is in any state with no less than T channels occupied, i.e.,

$$P_B = \sum_{i=T}^{C} P(i). \tag{3.16}$$

The handoff dropping probability P_D is the probability that the system is in the state with all C channels occupied, i.e.,

$$P_D = P(C). \tag{3.17}$$

It is obvious from (3.16) and (3.17) that $P_B \geq P_D$. Compared with the non-priority handoff scheme, the GC scheme gives a higher priority to handoff call requests than new call requests.

3.2   Heterogeneous Bandwidth Applications

Wireless multimedia networks support not only traditional voice applications, but also other applications such as data, audio and video, which have various bandwidth resource requirements. In such a heterogeneous application scenario, the assumption of each call requiring one channel made in Section 3.1 is not valid. Given the same state definition as the number of channels occupied in the cell, the state transition of a cell occurs not only between the immediate neighboring states any more. The arrival/departure of a call of other applications requiring r channels results in the state transition from current state x to the state $x + r/x - r$. An example system with three applications requiring one, two and three channels respectively is illustrated in Fig. 3.3.

The system shown in Fig. 3.3 cannot be described by a simple birth-death process, since the state transition probability $p(i, k) \neq 0$ for $|k - i| > 1$. Consequently, it is not straightforward to solve the set of equilibrium difference equations to get the steady state distribution as for the birth-death process.

However, by redefining the state, we can decompose the process into multiple birth-death processes . In general, let us assume that

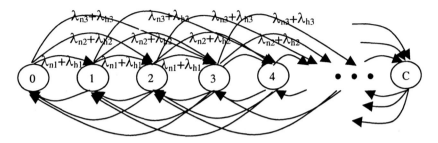

Figure 3.3: Illustration of the state transition diagram of a system with different channel requirements.

there are N different types of applications requiring different bandwidth of $r_1, r_2, ...r_N$ channels, respectively. Define the state $\mathbf{x}$ of a cell as an N-tuple vector $\mathbf{x} = (x_1, x_2, ...x_N)$, where $x_i(1 \leq i \leq N)$ is the number of calls belonging to the ith type of application in the cell. The state space then becomes

$$\Lambda = \left\{ \mathbf{x} = (x_1, x_2, ...x_N)| \sum_{i=1}^{N} x_i r_i \leq C \right\}. \qquad (3.18)$$

The following assumptions are made accordingly.

- New call requests of the ith type of application to the cell is a Poisson process with rate λ_{ni}

- Handoff requests of the ith type of application to the cell is a Poisson process with rate λ_{hi}

- The time for each call of the ith type of application staying in the cell is exponentially distributed with mean cell residence time $\frac{1}{\eta_i}$

- The arrivals of new and handoff call requests from each type of application are independent of each other.

- The arrival and the departure of each type of application are independent of other types of application.

Based on the above assumptions, each type of application can be modeled as a birth-death process. The system can then be viewed as an N-dimensional birth-death process with each dimension independent of others, and with the constraint of the state space described in (3.18). Let $P_i(x_i)$ be the marginal steady state distribution of the ith type of application. Generally, $P_i(x_i)$ can be expressed as the following form,

$$P_i(x_i) = \begin{cases} f_i(x_i)\dfrac{1}{\sum\limits_{k=0}^{\lfloor\frac{C}{r_i}\rfloor} f_i(k)}, & 0 \le x_i r_i \le C, \\ 0, & \text{otherwise,} \end{cases} \qquad (3.19)$$

where $\lfloor a \rfloor$ represents the largest integer number no greater than the real variable a. Then, the steady state distribution of the N-dimensional birth-death process, i.e., the joint probability of N dimensions $P(x_1, x_2, ...x_N)$, can be derived as a product form

$$P(\mathbf{x}) = P(x_1, x_2, ...x_N) = \begin{cases} \dfrac{\prod\limits_{i=1}^{N} f_i(x_i)}{\sum\limits_{\mathbf{x}\in\Lambda}\prod\limits_{i=1}^{N} f_i(x_i)}, & \mathbf{x} \in \Lambda, \\ 0, & \text{otherwise.} \end{cases} \qquad (3.20)$$

3.2.1   Non-priority Handoff Scheme

If the system with heterogeneous bandwidth applications uses the non-priority handoff scheme for each type of application, according to (3.9) and (3.10), the marginal steady state distribution of the ith dimension becomes

$$P_i(x_i) = \begin{cases} \dfrac{(\frac{\lambda_{ni}+\lambda_{hi}}{\eta_i})^{x_i}}{x_i!}\left(\sum\limits_{k=0}^{\lfloor\frac{C}{r_i}\rfloor}\dfrac{(\frac{\lambda_{ni}+\lambda_{hi}}{\eta_i})^k}{k!}\right)^{-1}, & 0 \le x_i r_i \le C, \\ 0, & \text{otherwise.} \end{cases}$$
$$(3.21)$$

Compared with (3.19), we have

$$f_i(x_i) = \frac{(\frac{\lambda_{ni}+\lambda_{hi}}{\eta_i})^{x_i}}{x_i!}. \qquad (3.22)$$

By substituting (3.22) into (3.20), we can get the steady state distribution as

$$P(\mathbf{x}) = P(x_1, x_2, ...x_N) = \begin{cases} \dfrac{\prod\limits_{i=1}^{N} \frac{(\frac{\lambda_{ni}+\lambda_{hi}}{\eta_i})^{x_i}}{x_i!}}{\sum\limits_{\mathbf{x}\in\Lambda}\prod\limits_{i=1}^{N} \frac{(\frac{\lambda_{ni}+\lambda_{hi}}{\eta_i})^{x_i}}{x_i!}}, & \mathbf{x} \in \Lambda, \\[10pt] 0, & \text{otherwise.} \end{cases}$$

(3.23)

Given the steady state distribution, the ith type of application requiring bandwidth of r_i channels has new call blocking probability P_{Bi} and handoff dropping probability P_{Di} as the probability that the system is in the states with more than $C - r_i$ channels occupied, namely,

$$P_{Bi} = P_{Di} = \sum_{\mathbf{x}\in\Lambda_i} P(\mathbf{x}),$$

(3.24)

where Λ_i is a subset of the state space Λ such that

$$\Lambda_i = \left\{ \mathbf{x} = (x_1, x_2, ...x_N) | C - r_i < \sum_{i=1}^{N} x_i r_i \leq C \right\}.$$

(3.25)

3.2.2 Priority Handoff Scheme

A wireless system with heterogeneous bandwidth requirements from different types of applications can also use the priority handoff scheme to give a preference to handoff requests than new call requests. Denote the threshold for the ith type of application requiring r_i channels by T_i, which is in the number of the ith type of calls. Then, T_i should satisfy the following constraint,

$$0 \leq T_i r_i \leq C, \qquad 1 \leq i \leq N.$$

(3.26)

For a system with state $\mathbf{x} = (x_1, x_2, ...x_N)$, a new call request of this type is admitted if and only if the number of this type of calls in the system is less than its threshold, i.e., $x_i < T_i$, while a handoff request of the ith type is admitted if there are still enough channels for this call, i.e., $\sum_{i=1}^{N} x_i r_i \leq C - r_i$.

For each individual type of call, the marginal steady state distribution of the ith dimension can be derived similarly to that of (3.14) and (3.15). Thus, the steady state distribution can be obtained from (3.20) as

$$
P(\mathbf{x}) = P(x_1, x_2, ... x_N) = \begin{cases} \dfrac{\prod\limits_{i=1}^{N} f_i(x_i)}{\sum\limits_{\mathbf{x} \in \Lambda} \prod\limits_{i=1}^{N} f_i(x_i)}, & \mathbf{x} \in \Lambda, \\ \\ 0, & \text{otherwise,} \end{cases} \tag{3.27}
$$

where

$$
f_i(x_i) = \begin{cases} \dfrac{(\frac{\lambda_{ni}+\lambda_{hi}}{\eta_i})^{x_i}}{x_i!} & 0 \leq x_i < T_i, \\ \\ \dfrac{(\frac{\lambda_{ni}+\lambda_{hi}}{\eta_i})^{T_i}(\frac{\lambda_{hi}}{\eta_i})^{x_i-T_i}}{x_i!} & T_i \leq x_i \leq \lfloor \frac{C}{r_i} \rfloor. \end{cases} \tag{3.28}
$$

Given the steady state distribution, the ith type of application requiring bandwidth of r_i channels has new call blocking probability P_{Bi}, which is the probability that the system in the states either with no less than T_i of the same type of calls or with less than r_i channels available. That is,

$$
P_{Bi} = \sum_{\mathbf{x} \in \Lambda_{Bi}} P(\mathbf{x}), \tag{3.29}
$$

where Λ_{Bi} is a subset of the state space Λ such that

$$
\Lambda_{Bi} = \left\{ \mathbf{x} | T_i < x_i \leq \lfloor \frac{C}{r_i} \rfloor \right\} \cup \left\{ \mathbf{x} | C - r_i < \sum_{i=1}^{N} x_i r_i \leq C \right\}. \tag{3.30}
$$

The handoff dropping probability P_{Di} is the probability that the system is in a state with more than $C - r_i$ channels occupied, i.e.

$$
P_{Di} = \sum_{\mathbf{x} \in \Lambda_{Di}} P(\mathbf{x}), \tag{3.31}
$$

where Λ_{Di} is a subset of the state space Λ such that

$$
\Lambda_{Di} = \left\{ \mathbf{x} | C - r_i < \sum_{i=1}^{N} x_i r_i \leq C \right\}. \tag{3.32}
$$

Apparently, we have $\Lambda_{Di} \subseteq \Lambda_{Bi}$, which leads to $P_{Di} \leq P_{Bi}$.

3.3   Rate-adaptive Applications

With the development of multimedia compression and coding technologies, more and more multimedia applications have the bandwidth adaptive capability. Wireless multimedia networks can exploit the adaptability of such applications to improve the efficiency of resource utilization.

A rate-adaptive application can adapt the required bandwidth of a media type to the available network resource. For example, a video phone call requiring a much higher bandwidth than the traditional voice phone call can be adapted to be a voice-only call if the network is heavily loaded. A web browsing application with video/audio/image can be configured to text-only when the network is congested.

For the same media type, adaptability can also be achieved by using a multilevel or a hierarchical coding scheme that enables it to adapt to different amounts of available bandwidth in the network. For example, voice applications can be encoded to a rate ranging from 8 KBps to 128 KBps by choosing appropriate encoding mechanisms or dynamically modifying the encoding parameters.

Video applications can be made rate-adaptive by using the layered coding system. For example, the MPEG-2 video/audio compression standard [2] defines different layers and profiles to achieve SNR and spatial scalability. The lowest layer (i.e. the base layer), consists of critical information for decoding video at the lowest visual quality. Additional layers provide increasing quality. Applications using this kind of codec can adapt to network resource availability by transmitting bit streams coded at different layers.

Another promising approach for adaptation is fine-granular scalable (FGS) coding systems. For example, the wavelet-based JPEG-2000 image coding standard [3] has an embedded coding stream. Instead of getting a few discrete coding rates provided by a layered coding system, continuous bit rates can be achieved by cutting a single coded bit stream at almost any bit. Better quality can be obtained by transmitting more bits in the bit stream. Similarly, MPEG-4 [4] is currently developing video coding with fine-granular scalability.

Generally, an adaptive application can choose its bandwidth from a set ranging from r_{min} to r_{max}, either discretely or continuously. As in previous sections, we consider two different schemes, non-priority and priority handoff, respectively, in the following two subsections.

3.3.1   Non-priority Handoff Scheme

The non-priority handoff scheme for rate-adaptive applications treats both new call and handoff call requests the same. When a request is made, if the network has enough resources available, the request can be admitted at r_{max}, which means that it is served at the highest quality. If the network traffic becomes heavier and r_{max} cannot be satisfied, the request of rate-adaptive applications can still be admitted at a degraded rate r_d, where r_d is in the set of bandwidth from which the application can choose and $r_{min} \leq r_d < r_{max}$, if r_d can be accommodated by the network. If the network becomes more congested such that it cannot provide r_{min} to the request, one or more existing rate-adaptive calls at r_{max} in the system can be degraded to lower rates. The released bandwidth resource can be used to admit the request. If the network is overloaded such that even if all existing rate-adaptive calls are at their r_{min}, there is still not enough resource available for the new call at its r_{min}, then the request is blocked. When a call leaves the system, the released bandwidth can also be used by other calls operating in the degraded mode to upgrade to a higher rate up to r_{max}.

There could be different bandwidth adaptation algorithms [5, 6, 7, 8, 9] to decide which calls should be degraded if necessary and how much r_d should be chosen upon an arrival of a request, as well as which calls should be upgraded and how much rate can be increased upon a departure. Since the bandwidth at which each call is conducting affects the end-to-end quality perceived by the user, different bandwidth adaptation algorithms result in different application-level QoS eventually. How to optimize the bandwidth adaptation algorithm according to application-level QoS measurement is another important issue. Here, we only consider connection-level QoS in terms of the call blocking/dropping probabilities. Thus, the system can be modeled as shown in Fig. 3.4 regardless of the detailed bandwidth adaptation algorithm.

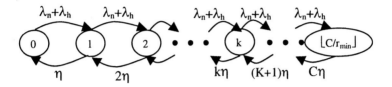

Figure 3.4: Illustration of the state transition diagram of a non-priority handoff system with rate-adaptive applications.

The same assumptions on channel capacity C, new call request rate λ_n, handoff request rate λ_h and channel holding time η are made as before. Also, we define the state of the system x to be the number of adaptive calls admitted in the system regardless the actual bandwidth used by each call. Then, the state space is

$$\Lambda = \left\{ x, | 0 \leq x \leq \lfloor \frac{C}{r_{min}} \rfloor \right\}.$$

Moreover, we call a state *degraded*, if the system has at least one ongoing call at a degraded rate r_d in this state. Consequently, we define two new metrics, i.e. *the system degradation probability* and *the call degradation probability*, respectively, to analyze the degradation. The system degradation probability, denoted by P_{sd}, is defined as the probability that the system is in degraded states. The call degradation probability, denoted by P_{cd}, is defined as the probability that a call request of adaptive application will experience degradation.

As shown in Fig. 3.4, it is again a M/M/C/C queuing model. Similar as that in Fig. 3.1, the steady state distribution can be derived as

$$P(k) = \begin{cases} \dfrac{(\frac{\lambda_n+\lambda_h}{\eta})^k}{k!} \dfrac{1}{\sum\limits_{i=0}^{\lfloor \frac{C}{r_{min}} \rfloor} \frac{(\frac{\lambda_n+\lambda_h}{\eta})^i}{i!}}, & 0 \leq k \leq \lfloor \frac{C}{r_{min}} \rfloor, \\ \\ 0, & \text{otherwise.} \end{cases} \qquad (3.33)$$

The new call blocking probability and the handoff dropping probability are the probability that the system is in the state with more than $(C - r_{min})$ channels occupied and all calls occupying the channels have already been degraded to r_{min}. That is,

$$P_B = P_D = \sum_{k \in \Lambda_{BD}} p(k),$$ (3.34)

where

$$\Lambda_{BD} = \{x, |C - r_{min} < xr_{min} \leq C\}$$ (3.35)

is a subset of Λ.

The system degradation probability P_{sd} is the probability that the system is in a degraded state and can be written as

$$P_{sd} = \sum_{k \in \Lambda_{ds}} p(k),$$ (3.36)

where

$$\Lambda_{ds} = \left\{ x \Big| \lfloor \frac{C}{r_{max}} \rfloor < x \leq \lfloor \frac{C}{r_{min}} \rfloor \right\}$$ (3.37)

is the state space of degraded states.

The call degradation probability P_{cd} can be calculated as

$$P_{cd} = \frac{E[N_d]}{E[N]},$$ (3.38)

where $E[N_d]$ is the expected number of calls at degraded rates in the system, and $E[N]$ is the expected total number of calls in the system. For each state x, let the number of degraded calls be denoted by x_d. Then, we have

$$E[N_d] = \sum_{x=0}^{\lfloor \frac{C}{r_{min}} \rfloor} x_d P(x),$$ (3.39)

and

$$E[N] = E[x] = \sum_{x=0}^{\lfloor \frac{C}{r_{min}} \rfloor} x P(x).$$ (3.40)

3.3.2 Priority Handoff Scheme

The priority handoff scheme for adaptive applications gives a higher priority to a handoff request over a new call request in the following way. When a new call request arrives, it is admitted if and only if the network can provide it with its full resource requirement r_{max}. Otherwise, the new call request is rejected. When a handoff request from an adaptive call is made to the cell, if the network has enough resources available, the call is handoffed into the cell at r_{max}. If the network traffic becomes heavier and r_{max} cannot be met, the handoff request of rate-adaptive applications can still be admitted at a degraded rate r_d, where $r_{min} \leq r_d < r_{max}$, if r_d can be accommodated by the network. If the network becomes more congested such that it cannot provide r_{min} to the request, one or more existing adaptive calls at r_{max} in the system can be degraded to lower rates. The released bandwidth resource can be used to admit the handoff request. If the network is overloaded, such that even if all existing rate-adaptive calls degraded to their r_{min}, there is still not enough resource available for the handoff request at its r_{min}. Then, the handoff fails and the adaptive call who makes the handoff request is dropped. When there is a call leaving the system, the released bandwidth can be used by other calls operating in the degraded mode to upgrade to a higher rate up to r_{max}.

To study the connection-level QoS, a cell using the priority handoff scheme for rate-adaptive applications can be modeled as shown in Fig. 3.5 regardless of the detailed bandwidth adaptation algorithm.

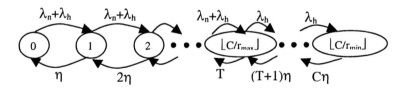

Figure 3.5: Illustration of the state transition diagram of a priority handoff system with rate-adaptive applications.

Compared with the non-priority handoff scheme for adaptive applications depicted in Fig. 3.4, the priority handoff scheme for adaptive applications admits only handoff requests when the system enters into degraded states. All the new call arrival request will be blocked. This is analogous to the GC scheme shown in Fig. 3.2, with threshold T set to $\lfloor \frac{C}{r_{max}} \rfloor$. Thus, the steady state distribution can be derived from results of the GC scheme as

$$
P(k) = \begin{cases}
\frac{(\frac{\lambda_n + \lambda_h}{\eta})^k}{k!} P(0), & 0 \le k < T, \\
\frac{(\frac{\lambda_n + \lambda_h}{\eta})^T (\frac{\lambda_h}{\eta})^{k-T}}{k!} P(0), & T \le k \le \lfloor \frac{C}{r_{min}} \rfloor, \\
0, & \text{otherwise,}
\end{cases}
\tag{3.41}
$$

where

$$
P(0) = \left(\sum_{i=0}^{T} \frac{(\frac{\lambda_n + \lambda_h}{\eta})^i}{i!} + \sum_{i=T+1}^{\lfloor \frac{C}{r_{min}} \rfloor} \frac{(\frac{\lambda_n + \lambda_h}{\eta})^T (\frac{\lambda_h}{\eta})^{i-T}}{i!} \right)^{-1},
\tag{3.42}
$$

and

$$
T = \lfloor \frac{C}{r_{max}} \rfloor.
\tag{3.43}
$$

Given the steady state distribution, the new call blocking probability P_B is the probability that the system is in any state with no less than T channels occupied, i.e.

$$
P_B = \sum_{i=T}^{\lfloor \frac{C}{r_{min}} \rfloor} P(i)
\tag{3.44}
$$

The handoff dropping probability P_D is the probability that the system is in states with more than $C - r_{min}$ channels occupied, and all calls occupying these channels have already been degraded to r_{min}. Thus, we have

$$
P_D = \sum_{k \in \Lambda_D} p(k),
\tag{3.45}
$$

where

$$
\Lambda_D = \{ x, | C - r_{min} < x r_{min} \le C \}.
\tag{3.46}
$$

The system degradation probability P_{sd} is the probability that the system is in degraded states. It is easy to obtain

$$P_{sd} = \sum_{k \in \Lambda_{sd}} p(k), \tag{3.47}$$

where

$$\Lambda_{sd} = \left\{ x, \lfloor \frac{C}{r_{max}} \rfloor < x \leq \lfloor \frac{C}{r_{min}} \rfloor \right\} \tag{3.48}$$

is the state space of degraded state.

The call degradation probability P_{cd} can be calculated in the same way as that given in Section 3.3.1 by using (3.38), (3.39) and (3.40).

3.4 Numerical Results

In this section, we give numerical results obtained for scenarios discussed in Sections 3.1, 3.2 and 3.3. For each scenario, the performance of connection-level QoS in terms of the new call blocking probability and the handoff dropping probability of a non-priority handoff scheme and a priority handoff scheme is compared. The degradation probability for adaptive applications is also evaluated.

3.4.1 Homogeneous Bandwidth Applications

Consider a cell with a capacity of $C = 36$ channels. The mean time of a call staying in the cell is 500 seconds, which means $\eta = 0.002$ call/sec. Assume the handoff rate λ_h is the half of new call arrival rate, i.e., $\lambda_h = 0.5\lambda_n$. To evaluate connection-level QoS under different traffic loads, the new call arrival rate λ_n varies from 0.02 to 0.5 seconds, i.e. the mean inter-arrival time of new call requests varies from 50 seconds to 2 seconds. The new call blocking probability and the handoff dropping probability for non-priority handoff scheme (denoted by NP) and priority handoff scheme (denoted by GC) with threshold $T = 32$ channels are compared in Fig. 3.6.

It is clear from Fig. 3.6 that, with the increase of λ_n, the traffic load of the cell becomes heavier. As a result, the new call blocking probability and the handoff dropping probability increase for both

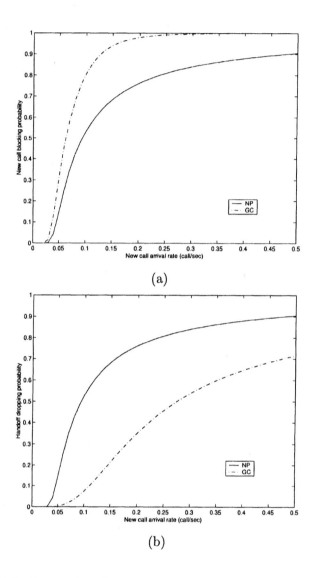

(a)

(b)

Figure 3.6: Comparison between non-priority and priority handoff schemes with homogeneous bandwidth applications ($T = 32$): (a) the new call blocking probability, (b) the handoff dropping probability.

schemes. It is not surprising that the new call blocking and the handoff dropping probabilities in the non-priority handoff scheme are the same for all λ_n, since the non-priority handoff scheme does not differentiate handoff requests from new call requests.

Fig. 3.6 (a) shows that the new call blocking probability of the non-priority handoff scheme is lower than that of the priority handoff scheme, while Fig. 3.6 (b) shows that the handoff dropping probability of the non-priority handoff scheme is higher than that of the priority handoff scheme. Thus, by giving a higher priority to a handoff than a new call, the priority handoff scheme reduces handoff dropping at the cost of blocking more new calls.

Threshold T plays an important role in the priority handoff scheme. To further evaluate the effect of different T, we fix the traffic load to $\lambda_n = 0.1$ call/sec while vary T from 1 to 36. Fig. 3.7 shows the trend of the new call blocking probability and the handoff dropping probability.

The tradeoff between the new call blocking probability and the handoff dropping probability is obvious in Fig. 3.7. With the increasing of threshold T, the new call blocking probability decreases while the handoff dropping probability increases.

3.4.2 Heterogeneous Bandwidth Applications

Consider a cell with the same capacity of $C = 36$ channels. In this section, we assume that there are three types of multimedia applications, e.g. voice, audio and video, with different bandwidth requirements of 1, 2 and 4 channels, respectively. The mean time of a call of any type of applications staying in the cell is 500 seconds. The new call requests by each type of application arrive independently with the same Poisson arrival rate, $\lambda_{n1} = \lambda_{n2} = \lambda_{n3}$. Thus, the total new call arrives at the cell with a Poisson arrival rate of $\lambda_n = 3\lambda_{ni}, (i = 1, 2, 3)$. The handoff arrival rate of each type of applications is set to be a portion of its new call arrival rate. Since the type of call with a higher bandwidth requirement tends to have a higher new call blocking probability, we set $\lambda_{h1} = 0.5\lambda_{n1}$, $\lambda_{h2} = 0.2\lambda_{n2}$, and $\lambda_{h3} = 0.1\lambda_{n3}$, respectively. For the priority handoff scheme, thresholds of each type of calls T_i are set to $T_i = \frac{C}{3r_i}$, such that the

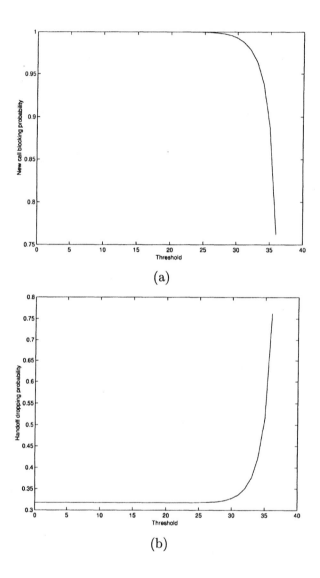

Figure 3.7: The effect of T on the priority handoff scheme with homogeneous bandwidth applications: (a) the new call blocking probability, (b) the handoff dropping probability.

total C channels are shared equally by new calls from three types of calls. The new call blocking probability and the handoff dropping probability for r1, r2, and r3 applications are shown in Figs. 3.8, 3.9, and 3.10, respectively.

As expected, the handoff dropping probabilities of all three types of calls under priority handoff (denoted by GC in these figures) are lower than those under non-priority handoff (denoted by NP in the figures). As a tradeoff, the new call blocking probabilities of r1 and r2 calls under priority handoff are higher than those under non-priority handoff. The result of the new call blocking probability of r3 calls shown in Fig. 3.10(b) is interesting. When the traffic load is relatively low, NP has a lower r3 new call blocking probability while, when the traffic load becomes heavier, GC has a lower r3 new call blocking probability. This can be explained by comparing new call blocking probabilities of three types of calls under each scheme. The NP scheme tends to block more higher-rate calls than lower-rate calls, while the GC scheme improves the fairness among different type of calls by threshold setting. As a result, when the traffic from all three types of calls becomes heavy in the system, the GC scheme blocks more r1 and r2 new call requests, and the resource is used to lower r3 new call blocking and handoff droppings.

3.4.3 Rate-adaptive Applications

In this section, we calculate connection-level QoS for rate-adaptive applications. Assume an adaptive application requires $r_{max} = 4$ channels at its highest layer quality. In the case of network congestion, it can tolerate $r_{min} = 1$ channel at its basic layer quality. Other assumptions about the traffic configuration are the same as those made in Section 3.4.1. First, we compare adaptive and non-adaptive applications under different handoff schemes to show the advantage of adaptive applications. Then, we compare two different handoff schemes for adaptive applications.

Non-priority Handoff

The above mentioned adaptive applications are compared with non-adaptive applications requiring 4 channels under non-priority handoff

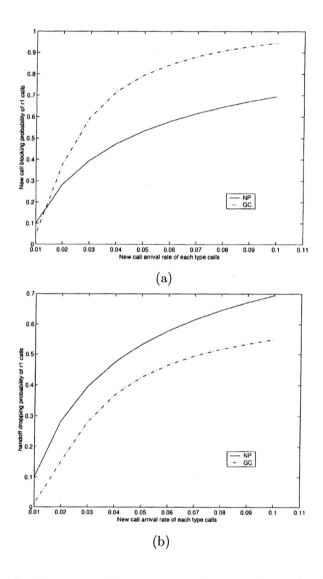

(a)

(b)

Figure 3.8: Comparison between non-priority and priority handoff schemes with heterogeneous bandwidth applications - r1 calls: (a) the new call blocking probability, (b) the handoff dropping probability.

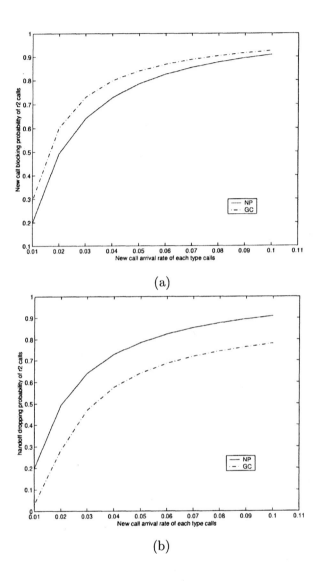

(a)

(b)

Figure 3.9: Comparison between non-priority and priority handoff schemes with heterogeneous bandwidth applications - r2 calls: (a) the new call blocking probability, (b) the handoff dropping probability.

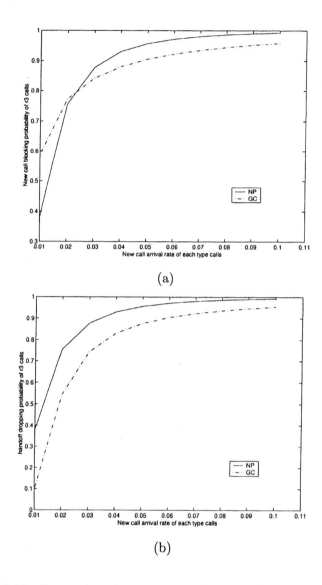

Figure 3.10: Comparison between non-priority and priority handoff schemes with heterogeneous bandwidth applications - r3 calls: (a) the new call blocking probability, (b) the handoff dropping probability.

in Fig. 3.11, where the new call blocking probability and the handoff dropping probability are given.

Fig. 3.11 demonstrates that by deploying the rate adaptability of multimedia applications, the system achieves a much lower new call blocking probability and a lower handoff dropping probability. Clearly, this improvement on connection-level QoS is at the expense of application-level QoS degradation. To show the extent to which the system and the application could fall into a degraded state, Fig. 3.12 plots the system degradation probability and the call degradation probability.

As given in Fig. 3.12, adaptive applications experience a higher degradation probability with the increase of the traffic load.

Priority Handoff

Furthermore, we compare adaptive with non-adaptive applications under priority handoff. For the same adaptive and non-adaptive applications considered in Section 3.4.3, we assume that the non-adaptive application chooses a threshold to be 32 channels, i.e., $T = 8$ calls. Fig. 3.13 shows the comparison of the new call blocking probability and the handoff dropping probability.

As shown in Fig. 3.13, the new call blocking probability of adaptive applications is close to that of non-adaptive applications with a slight increase under a heavy traffic load. This is because of the relatively lower threshold for adaptive calls than that for non-adaptive calls used in our example. On the other hand, the handoff blocking probability is greatly reduced for adaptive applications than that for non-adaptive applications. The gain on connection-level QoS is again at the cost of application-level quality degradation. Fig. 3.14 shows the system degradation probability and the call degradation probability.

Similar to that under non-priority handoff, Fig. 3.14 shows that adaptive applications experience a higher degradation probability with an increase of the traffic load under priority handoff.

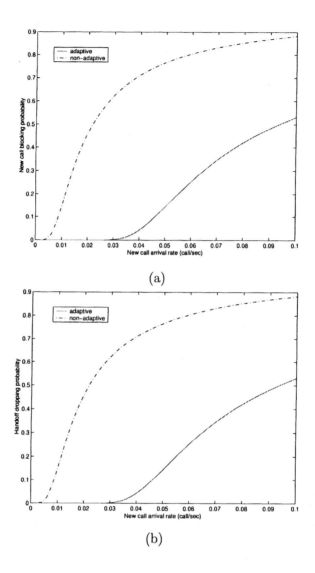

Figure 3.11: Comparison between adaptive and non-adaptive applications under non-priority handoff: (a) the new call blocking probability, (b) the handoff dropping probability.

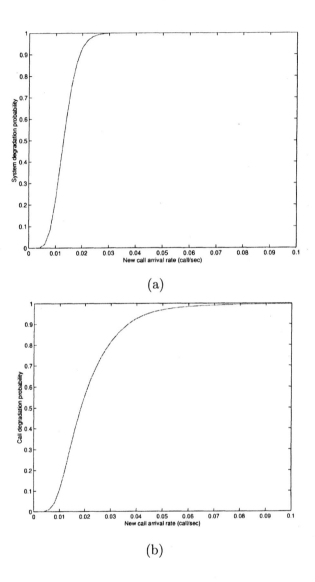

(a)

(b)

Figure 3.12: Degradation probabilities of adaptive applications under the non-priority scheme: (a) the system degradation probability, (b) the call degradation probability.

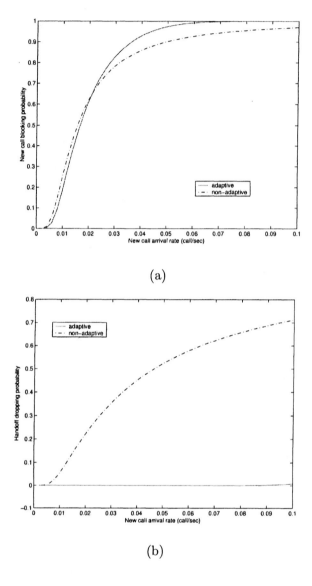

(a)

(b)

Figure 3.13: Performance comparison between adaptive and non-adaptive applications under priority handoff: (a) the new call blocking probability, (b) the handoff dropping probability.

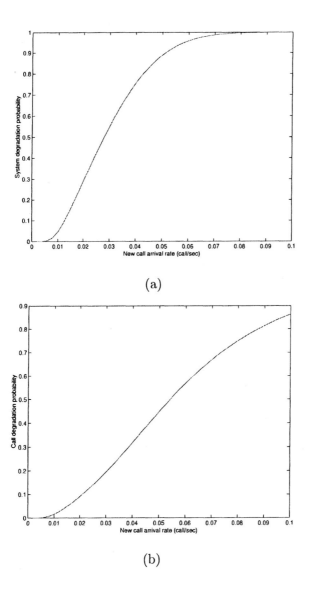

(a)

(b)

Figure 3.14: The degradation probabilities of adaptive applications under priority handoff: (a) the system degradation probability, (b) the call degradation probability.

Comparison between Non-priority Handoff and Priority Handoff

Finally, we compare the non-priority handoff scheme with the priority handoff scheme for adaptive applications. The new call blocking probability and the handoff dropping probability are compared in Fig. 3.15, while the system degradation probability and the call degradation probability are compared in Fig. 3.16.

Figs. 3.15(a) and (b) show the tradeoff between the new call blocking probability and the handoff dropping probability by using non-priority and priority handoff schemes for adaptive applications, which is similar to that for non-adaptive applications in Section 3.4.1. Moreover, from Figs. 3.16(a) and (b), we can see that by giving higher priority to handoff calls than new call requests, the priority handoff scheme for adaptive applications reduces the probabilities of the system and the individual call in degradation states.

3.5   Conclusion and Future Work

In this chapter, we studied mathematical models of wireless multimedia networks to analyze connection-level QoS in terms of the new call blocking probability and the handoff dropping probability. We consider two different schemes, non-priority handoff and priority handoff (more specifically, GC), which are simple and typical network mechanisms used in wireless communication networks. First, a basic queuing model was used to analyze traditional wireless communication networks with voice applications only. Then, the basic model was extended to wireless multimedia networks with different types of applications requiring different bandwidth resources, such as voice, audio and video applications. Finally, we incorporated the rate-adaptive applications to the model analysis. The gain on connection-level QoS and the degradation on application-level QoS by deploying the rate-adaptability are analyzed.

Based on the model analysis, we make the following conclusions.

- For a system with homogeneous bandwidth applications, priority handoff can achieve a much lower handoff dropping probability than non-priority handoff at the cost of a higher new

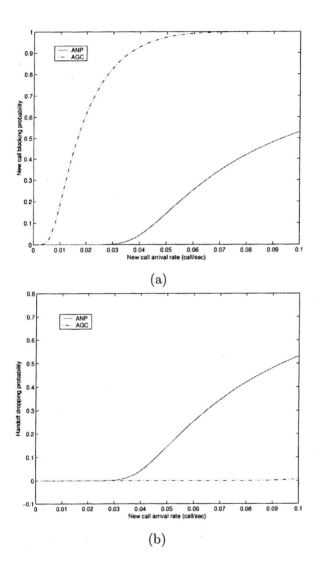

Figure 3.15: Performance comparison between non-priority and priority handoff schemes for adaptive applications: (a) the new call blocking probability, (b) the handoff dropping probability.

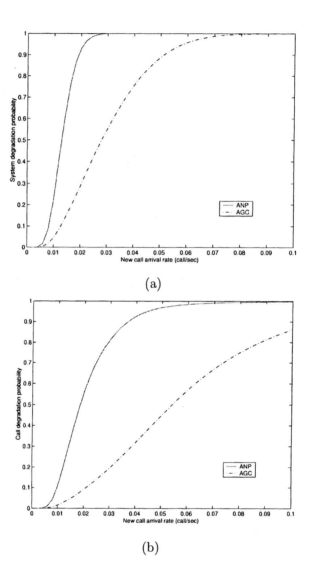

(a)

(b)

Figure 3.16: Performance comparison between non-priority and priority handoff schemes for adaptive applications: (a) the system degradation probability, (b) the call degradation probability.

call blocking probability.

- For a system with heterogeneous bandwidth applications, priority handoff can achieve a lower handoff dropping probability and better fairness in traffic loads among different applications.

- Adaptive multimedia applications achieves a lower new call blocking probability and a lower handoff dropping probability at the cost of application-level QoS degradation compared to non-adaptive applications.

- For adaptive applications, priority handoff can achieve a lower handoff dropping probability, as well as a lower system degradation and a lower call degradation probability at the cost of a higher new call blocking probability than non-priority handoff.

As a future work direction in mathematical model analysis, more sophisticated network mechanisms can be investigated by using more complicated models. However, there are always limitations on pure mathematical analysis. Throughout model analysis in this chapter, some assumptions were made for tractability with a mathematical method. In a real-world wireless network, these assumptions may not be valid. For example, it is assumed that handoff and new call arrivals are independent Poisson processes in our model analysis, while in real-world wireless networks, handoff arrivals are related to new call arrivals, the mew call blocking probability and user's mobility characteristics. Moreover, the exponential distribution assumed for the time of calls staying in a cell may not be suitable to describe the statistical behavior of wireless multimedia applications. As a result, the models analyzed in this chapter provide only a simplified reference to real-world wireless multimedia network scenarios, which are usually too complicated to model and analyze. To evaluate the actual performance of network protocols and mechanisms in real scenarios, computer simulations using network simulators should be conducted. In the next chapter, we will conduct both mathematical model analysis and simulations to evaluate our proposed adaptive QoS management scheme for wireless multimedia networks.

BIBLIOGRAPHY

[1] L. Klennrock, *Queueing systems Volume 1: Theory*. John Wiley & Sons, Inc., 1975.

[2] ISO/IEC JTC1/SC29/WG11, "Generic coding of moving pictures and associated audio information," ISO/IEC International Standard 13818, Nov. 1994.

[3] ISO/IEC JTC1/SC29/WG1, "JPEG2000 part 1 final committee draft version 1.0," ISO/IEC International Standard N1646R, Mar. 2000.

[4] ISO/IEC JTC1/SC29/WG11, "Overview of the MPEG-4 standard," ISO/IEC N3747, Oct. 2000.

[5] T. Kwon, I. Park, Y. Choi, and S. Das, "Bandwidth adaption algorithms with multi-objectives for adaptive multimedia services in mobile networks," in *Proceedings of the second ACM international workshop on Wireless mobile multimedia*, 1999, pp. 51 – 59.

[6] T. Kwon, J. Choi, Y. Choi, and S. Das, "Near optimal bandwidth adaptation algorithm for adaptive multimedia services in wireless/mobile networks," in *Proc. of the 50th IEEE Vehicular Technology Conference*, 1999, vol. 2, pp. 874 –878.

[7] V. Bharghavan, K.-W. Lee, S. Lu, S. Ha, J.-R. Li, and D. Dwyer, "The TIMELY adaptive resource management architecture," *IEEE Personal Communications Magazine*, vol. 5, pp. 20-31, 1998.

[8] S. K. Sen and S. Das, "Quality-of-service degradation strategies in multimedia wireless networks," in *Proc. of IEEE Vehicular Technology Conference (VTC'98)*, May 1998.

[9] K. Lee, "Adaptive network support for mobile multimedia," in *Proc. of ACM Mobicom'95*, Nov. 1995, pp. 62–74.

Chapter 4

WIRELESS NETWORKS WITH ADAPTIVE RESOURCE MANAGEMENT

Supporting multimedia services in wireless networks presents more challenges than in a wired network due to the limited bandwidth resources and the highly variable environment. Furthermore, user's mobility makes it more complicated to achieve the desired QoS level of multimedia services. For these reasons, the best a system can do is to provide different QoS according to different service requests from end users under the constraint of limited and varying bandwidth resources. This has been one of the key elements in the design of the existing network infrastructure providing QoS to multimedia applications, such as ATM, IP InteServ, and DiffServ. Motivated by the same concept, we investigate adaptive resource allocation for multimedia services in a wireless network environment. Both user mobility and multimedia traffic characteristics are taken into consideration in the development of the proposed adaptive scheme.

4.1 Adaptive Resource Management System

The proposed system of providing multimedia QoS services is illustrated in Fig. 4.1. The components of the proposed system have the following features.

- Service model: The proposed service model considers both user's mobility and multimedia traffic characteristics in order to provide suitable QoS to different classes of applications under the competing environment of limited bandwidth in wireless multimedia networks.

- Application profile: Application profiles are mapped to the service model by using different forms of traffic specifications.

- Resource allocation: Resources are allocated to different service classes adaptively by employing dynamic call admission control and resource reservation schemes according to application profiles.

- Adaptation: To maintain a specified QoS level, our system is adaptive not only to different service QoS requirements but also to network conditions. It exploits the rate-adaptive feature of multimedia applications to further improve the efficiency of resource utilization.

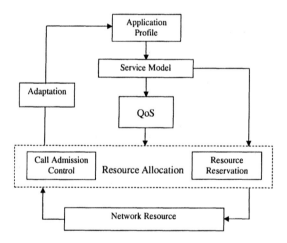

Figure 4.1: The proposed system for QoS provisioning.

Details of each component of the proposed system are described in the following subsections.

4.1.1 Service Model

Service model describes a set of offered services by the network. To provide a QoS to each user tailored to its requirements, network offers a menu of explicitly defined services. Applications choose the desired service from this menu according to their needs. In many QoS-aware networks, an appropriate service model is the foundation of QoS provisioning [1]. For example, ATM service model, which is designed for inherent QoS support, is composed of constant bit rate (CBR), real-time variable bit rate (rt-VBR), non-real-time variable bit rate (nrt-VBR), available bit rate (ABR), and unspecified bit rate (UBR) services [2]. The Internet Protocol (IP) was originally designed for the datagram network to provide the best-effort service without QoS support. To support the integrated multimedia service, its service model has been extended to include guaranteed and predictive services for real-time services and traditional best-effort for non-real-time services [3]. The Two-bit Differentiated Services Architecture (TDSA) [4] contains premium, assured, and best-effort service classes.

In a wireless multimedia network environment, an appropriate new service model is vital for QoS support due to limited resources. Researchers have tried to address the problem of providing QoS in wireless networks by considering different kinds of services. For example, Oliveira *et al.* [5] distinguished real-time and non real-time services without explicit service models. Talukdar *et al.* [6] proposed a three-class service model extended from that of the IP InteServ directly for the wireless integrated packet network. However, a comprehensive service model that can accommodate more realistic multimedia services is not yet commonly used.

In order to design a service model for a certain network infrastructure, primitive QoS requirements of multimedia applications to the network should be used to classify services. In ATM networks, the major traffic demands on networks are delay/delay variation (or jitter) and bandwidth. In terms of delay and delay variation sensitivity of different applications, CBR and rt-VBR service classes requires stringent delay/delay variation bound; while nrt-VBR, ABR, and UBR service classes do not have a tight constraint on delay and delay variations. In terms of the bandwidth demand, CBR service class requires a fixed data rate continuously available during the connection lifetime. VBR service classes has bursty source transmitting at a rate that varies with time. ABR can adjust its rate according to network bandwidth availability, and UBR just requires whatever resource left in the network. The IP InteServ service model identifies delay as the primitive QoS parameter for service classification. A guaranteed service provides a firm end-to-end delay bound to those delay-intolerant real-time applications, while the predictive service provides delay-tolerant real-time applications with a loose bound on delay by incorporating predictions about the aggregate traffic load. The bound could be occasionally violated if predictions are wrong. The best-effort service provides no delay bound to non-real-time applications. IP DiffServ considers the combination of delay and loss requirements to build up its two-bit service model. Compared to the best-effort service which does not have any delay or loss bound, applications requiring a more stringent delay bound are served as the premium service, which is based on *expedited-forwarding* (EF) *per-hop behavior* (PHB); while applications requiring a less stringent delay bound but more critical on packet loss receive the assured service based on *assured-forwarding* (AF) PHB. In the design of the service model for wireless multimedia networks, we follow the same design principle, i.e. consider the application requirements to the network first, categorize applications accordingly, and then propose service classes that meet these requirements.

Connection-level QoS Service Model

Since the connection-level QoS, in terms of the new call blocking probability and the handoff dropping probability, is an important issue for wireless networks, it is reasonable to choose from these two QoS measurement parameters for the purpose of classifying services. We select the handoff dropping probability as one of the primitive QoS requirements made by the application to the network to classify the applications due to the following reasons. First, handoff dropping is more intolerable to a mobile user than new call blocking.

Consequently, from user's viewpoint, the handoff dropping probability should have a more significant impact on the overall connection-level QoS, as well as the packet-level QoS. Second, the selected QoS measurement should be able to differentiate the service commitment made by the network on how to serve an admitted application rather than that of whether to serve an application.

Having the handoff dropping probability as the primitive QoS requirement, we can categorize applications to the following three categories.

- Handoff-guaranteed service: It represents real-time applications that require absolute continuity, i.e. no handoff dropping is permitted before the call is completed.

- Handoff-prioritized service: It represents real-time applications that can tolerate a reasonably low handoff dropping probability, but still demand a higher priority for handoff calls.

- Handoff-undeclared service: It represents real-time applications that do not demand the handoff priority, but require access to the network on the spot. The handoff request from this class has the lowest priority, and receives the best-effort kind of service in terms of handoff. If a handoff occurs to an application of this class when the network has a light load, it will be successfully handed off. Otherwise, it will be dropped first.

There could be different ways to decide into which class an particular application request should be classified. In our design, we consider the mapping of user's mobility such as the moving range and/or speed to the service model. First, mobile users with high mobility move fast within a large area across many cells, e.g. a moving vehicle on the highway. Handoff occurs frequently in this case. Consequently, the chance that the call being forced terminated before completion would be high even if the handoff dropping probability in each individual cell is relatively low. Therefore, users with high mobility would like to subscribe their applications to handoff-guaranteed service. Second, some mobile hosts have moderate mobility, as they move relatively slowly and within a smaller area around their current position, e.g. moving vehicles on the local roads. Hand-off occurs at a relatively lower rate with certain probability distributions in this case. It is neither necessary nor feasible to guarantee no handoff dropping. A sufficiently low handoff dropping probability are suitable for these applications. Thus, users with moderate mobility subscribe their applications to handoff-prioritized service. Finally, some users need wireless access, but are either stationary or have little movement such as users in an airport lobby waiting for a flight or pedestrians walking around a shopping mall. In these cases, almost no handoff is expected. Obviously, the handoff-undeclared service is suitable for users with low mobility to subscribe.

There could be other issues to be considered for service classification. For example, the priority of an application. Some calls, such as emergency rescue or business transactions, cannot be dropped before completion. These applications can require the guaranteed service. Some other calls, such as normal conversations, are not so critical in being dropped yet preferably not being interrupted. These applications may require the prioritized service. Since the priority is usually associated with pricing, there still could be some applications to be served at the lowest price if they do not mind being dropped during handoff. These applications can require the best-effort service.

Packet-level QoS Service Model

With the explosive increase of the wireless communication demand, the next-generation wireless networks will employ the packet-switching technology to support multimedia services. It is envisioned that the wireless system beyond 3G will evolve into an all-IP system, integrating advantages of the Internet and the mobile systems.

Packet-switched networks can take advantage of a higher degree of multiplexing among services, thus achieving better utilization of radio resources and higher efficiency of the total system. However, packets of certain service classes may experience varying delay, delay jitter and loss, which may have a great impact on the perceived quality of the service by the end user. The QoS guarantee for multimedia services in the packet-based next-generation wireless communication networks presents a great challenge at both the connection-level and the packet-level.

In packet-switched wireless networks, besides the handoff dropping probability, the delay requirement and traffic characteristics of applications should also be considered in the design of our service model in the context of wireless multimedia networks.

First, multimedia applications are classified into real-time and non real-time applications according to their delay requirements. Real-time applications, such as video conferencing, interactive games, and phone conversations have a stringent constraint on delay and delay variations. On the other hand, non real-time applications such as telnet, FTP, and email can tolerate relatively larger delay and delay variations. These applications are also called elastic since they are able to stretch gracefully when delay increases. Furthermore, according to the characteristics, real-time applications are further categorized to *constant bit rate* (CBR) and *variable bit rate* (VBR) according to the traffic characteristics. Based on the above discussion, we consider three types of service classes as defined below.

- Constant Bit Rate (CBR): The CBR traffic is for real-time applications that require constant bit rate available during the entire lifetime of connection. It has the highest priority. In this case, the packet-level QoS is guaranteed as long as the required bandwidth is guaranteed.

- Variable Bit Rate (VBR): The VBR traffic is for real-time applications that have a varying bit rate with burst at peak rates. Since this kind of applications can tolerate a small amount of delay or packet loss, it is possible to admit a VBR flow even if its peak rate cannot be accommodated. Therefore, the VBR traffic is more efficient in resource utilization through multiplexing. In this context, we are able to conduct the trade-off between packet-level and connection-level QoS.

- Unspecified Bit Rate (UBR): The UBR traffic is for non-real-time applications that do not have any requirement on the packet-level QoS such as delay and packet loss. It is equivalent to the best-effort class in the wired Internet, which uses whatever resource not being used in the network. Therefore, it make more efficient use of the scarce resource in wireless networks and improves system utilization.

4.1.2 Application Profile

The application profile is described by a set of parameters, such as the traffic pattern, the bandwidth requirement, the maximum delay, and the packet loss rate, etc. These parameters characterize service classes in the service model and are used by the network to take proper action for each service class.

For connection-level QoS service model, the application profile includes the required handoff dropping probability and the mobility information. For the handoff-guaranteed service, the target handoff dropping probability P_{DG} should be 0. For the handoff-prioritized service, the target handoff dropping probability P_{DP_tar} is bounded by $0 < P_{DP_{tar}} < 1$. Moreover, an application requesting the handoff-guaranteed service should provide its required bandwidth and mobility information in order to get the desired resources when it moves to other cells. From the mobility information, the network could predict the cells that the mobile is going to visit during its lifetime. For the handoff-undeclared service, it is implied that the handoff probability is upper bounded by 1, and it remains in the originating cell.

For packet-level QoS service model, the CBR traffic specifies its required bandwidth to guarantee the delay and the loss performance. The VBR traffic describes its profile with a set of parameters for its traffic model such as the peak rate, the mean rate and the state transition probability. The packet-level QoS metrics, such as the delay and/or loss constraints, are also necessary. For the UBR traffic, there is neither minimum bandwidth requirement nor delay/loss constraints.

4.1.3 Resource Allocation

To implement the service model described in the above section and achieve appropriate QoS for different service classes, a network should use different strategies to allocate its resources to calls belonging to different service classes according to their profile For the connection-level QoS service model, it is not sufficient to consider local resource allocation. Since mobility plays an important role in connection-level QoS provisioning, the resource within a larger range should be allocated globally through reservation.

The resource reservation mechanism can be broadly classified as *isolated reservation* and *shared reservation*. Isolated reservation reserves resources for each call separately, and the reserved resource can only be used by this call. This could result in lower resource utilization when the reserved resources are not actually being used. However, resources are guaranteed for each admitted call. On the other hand, shared reservation reserves aggregate resources for a group of calls. Compared to isolated reservation, higher utilization can be achieved in this case due to statistical reservation. However, it does not guarantee a sufficient amount of resources for each call.

A different resource reservation scheme, as explained below, is used for each service class to provide appropriate connection-level QoS to the corresponding applications.

- For the handoff-guaranteed service class, it is necessary to reserve resources in other cells which the mobile host may visit, indicated by its mobility information in the application profile. Isolated resource reservation is employed explicitly upon the admission of calls belonging to this class. Call admission control is simply based on whether resources are reserved successfully.

- For the handoff-prioritized service class, it is not necessary to make isolated reservation for each call. Instead, shared reservation is applied to this service class through a priority handoff scheme. To maintain a reasonably low target handoff dropping rate, we design a measurement-based algorithm that dynamically adjusts the threshold in the guard channel scheme. This algorithm will be described in the Section 5.1.

- For the handoff-undeclared classes, resources in other cells than the originating cell are not necessarily to be reserved, since handoff is not expected for these classes. The admission of these classes of calls is only based on resource availability in the current cell. This scheme is equivalent to the non-priority scheme described in Chapter 3. From results shown in Section 3.4, calls admitted to these classes could experience a high dropping probability if they attempt to make handoff requests.

For the packet-level QoS service model, resource allocation is different for real-time and non-real-time traffics. To achieve the desired QoS for a real-time application, it is usually necessary to maintain a minimum bandwidth during its lifetime. We adopt the virtual circuit concept to establish the connection for a real-time application including the CBR and the VBR traffic requests. The setup of a connection requires call admission control and resource reservation to prevent network congestion and dropping of on-going calls. The admitted CBR traffic is allocated its required bandwidth during its lifetime such that its packet-level QoS is guaranteed. The admitted VBR traffic shares the available resource in the manner that its packet-level QoS is guaranteed statistically. For non-real-time applications, i.e. the UBR traffic, we use the best-effort service adopted in traditional IP networks. Data from these applications can be stored at a network node, such as the base station or a mobile terminal. Whenever the network has spare resources, unused by real-time applications, these applications will be serviced under an appropriate scheduling algorithm. To improve resource utilization of the entire network, we may also use resources reserved for real-time applications but not yet being in use to carry non-real-time data. No call admission control or resource reservation is required here.

As discussed in Section 4.1.2, the application profile of each service class is used by the network to allocate adequate resources in order to provide desired QoS for each class. However, if an admitted call does not conform to its application profile, the network may not have adequate resources to maintain the committed service for every on-going call. Thus, to prevent bad calls violating their application profile from degrading QoS of other conforming calls, a policing mechanism should be enforced by the network. As an example, for the connection-level QoS service model, if a call of the handoff-guaranteed service class tries to enter a cell not covered by its mobility pattern it violates its service specifications. The policing mechanism will release the resources reserved for this call in all the cells included in its application profile and treat it as a handoff-prioritized call. If this call still needs the handoff-guaranteed service, it has to make a new request with a new application profile.

4.1.4   Rate-adaptive Applications

The wireless network is a highly variable environment. To maintain a specified QoS level, a wireless system has to adapt to varying conditions when the wireless link fluctuates or degrades. By using rate-adaptive features of many multimedia applications, our proposed resource allocation scheme can be adaptive to network conditions.

With the development of multimedia compression and coding technologies, more and more real-time applications such as video and audio codecs can be made rate-adaptive. For example, voice applications can be encoded at a rate ranging from 2 KBps to 128 KBps by choosing appropriate encoding mechanisms or dynamically modifying the encoding parameters. Video applications can be made rate-adaptive by using a layered coding method. For example, the MPEG-2 video/audio compression standard [7] defines different layers and profiles to achieve SNR and spatial scalability. The lowest layer (i.e. the base layer) consists of critical information for decoding the image sequence at its lowest visual quality. Additional layers provide increasing quality. Applications using this kind of codecs can adapt to network resource availability by transmitting bit streams coded at different layers. Another promising approach for adaptation is the use of embedded coding schemes, such as the wavelet-based JPEG-2000 image coding standard [8]. Instead of a few discrete coding rates provided by a layered coding scheme, continuous bit rates can be achieved by cutting a single coded bit stream at almost any bit. Better quality can be obtained by transmitting more bits in the bit stream. Similarly, MPEG-4 [9], which is the new generation multimedia communication coding standard, has the fine-granular scalability (FGS) mode.

For a rate-adaptive applications making a connection request to the network, they specify the range of bandwidths required to be supported by the network as $[MinBW, MaxBW]$, where $MinBW$ and $MaxBW$ denote the minimum and maximum bandwidth requirements, respectively. Adaptation first takes place while admitting a new call. If the network has enough resources available, the request can be admitted at $MaxBW$, which means that it is served at the highest quality. If the network becomes congested, a rate-adaptive application can degrade to a lower rate with degraded quality of service. If the network is overloaded and $MinBW$ cannot be satisfied, the call is blocked. Adaptation also takes place at the time of handoffs. A rate-adaptive connection admitted at $MaxBW$ could be handed off at a lower rate if the cell it is entering is heavily loaded. On the other hand, a call admitted at $MinBW$ could be upgraded to a higher rate if the cell it is going to enter is under-utilized. We use the rate adaptation for a call only upon its admission and handoffs, because frequent changes in the quality are not desirable for audio/visual applications. For example, watching video with rapidly changing quality back-and-forth is much more annoying than just watching it with constant but lower quality.

4.2   Analysis for Connection-level QoS Service Model

To further explain connection-level QoS provisioning for the real-time service class, we present mathematical analysis of our resource allocation scheme for the connection-level QoS service model in this section. First, the handoff-guaranteed service class is modelled individually, and then the handoff-prioritized service class is added into the model. Finally, the system is modelled by considering all the three service classes, *i.e.* handoff-guaranteed, handoff-prioritized, and handoff-undeclared services.

4.2.1   System with Handoff-guaranteed Service Class

Let us consider a single cell with a fixed amount of bandwidth (a total of C channels of the same bandwidth), we derive the model for the handoff-guaranteed service. Let n denote the number of cells a call is going to enter while travelling along a certain path during its life time. Upon admission, the channel resource is required to be reserved in each of the n cells, which it is going to enter. We define random variable T_i as the channel occupation time of the call in the ith $(1 \leq i \leq n)$ cell, which is equal to the time from the admission of the call till its leaving the ith cell. Thus, T_i equals to the sum of i independent random variables of the cell residence time t_j in cell j, i.e.

$$T_i = \sum_{j=1}^{i} t_j, \quad 1 \leq i \leq n. \tag{4.1}$$

Assuming that the cell residence time t_j is exponentially distributed with the mean cell residence time $\frac{1}{\eta}$, it can be shown that T_i has an i-stage Erlangian distribution with the density function

$$f(T_i) = \frac{(\eta T_i)^{i-1}}{(i-1)!} \eta \exp^{-\eta T_i}, \quad T_i \geq 0, i \geq 1. \tag{4.2}$$

Here, the stage parameter i decides the shape and moments of the Erlangian distribution. For $i = 1$, this is the same as the exponential distribution with the density function

$$f(T_1) = \eta \exp^{-\eta T_1}, \quad T_1 \geq 0. \tag{4.3}$$

The first cell is the originating cell where the resource is only occupied for the period of the cell residence time, i.e. $T_1 = t_1$.

It is assumed that the maximum number of cells a handoff-guaranteed call can traverse is N. Then, a cell with such a service class can be modeled as an N-dimensional model, where dimension m $(1 \leq m \leq N)$ is for all the calls entering this cell as their mth cell. Thus, the mth dimension is an $M/E_r/C/C$ queuing model with the Poisson arrival and the m-stage Erlang departure $(1 \leq m \leq N)$. Let the number of calls in the mth dimension be x_{Gm}. The state is the N-tuple vector $\mathbf{x} = (x_{G1}, x_{G2}, ...x_{GN})$ and the state space is $\Lambda = \{\mathbf{x}, |0 \leq \sum_{m=1}^{N} x_{Gm} \leq C\}$.

Given the steady state distribution $P(\mathbf{x})$ of the model, the blocking probability in each dimension is

$$P_{Bm} = \sum_{\mathbf{x} \in \Lambda, \sum_{m=1}^{N} x_{Gm} = C} P(\mathbf{x}). \tag{4.4}$$

The new call blocking probability $P_{BG}(n)$ $(1 \leq n \leq N)$ of a handoff-guaranteed call, which would visit n cells during its life time, can be computed as the probability that at least one of the n cells has no available channel for the new call request. Under the assumption of a unified cell load in all cells, we have

$$P_{BG}(n) = 1 - \prod_{m=1}^{n} (1 - P_{Bm}) = 1 - (1 - P_{Bm})^n. \qquad (4.5)$$

Since handoff is guaranteed not to be dropped for this type of calls, the handoff dropping probability $P_{DG} = 0$.

4.2.2   System with Handoff-guaranteed and Handoff-prioritized Service Classes

Let us consider a cell with C channels offering both handoff-guaranteed and handoff-prioritized service classes described above. The system can be analyzed by using a multi-dimensional model as follows. The state of a system is the vector $\mathbf{x} = \{x_G, x_P\}$, where x_G and x_P are the numbers of occupied (being used or reserved) channels by handoff-guaranteed and handoff-prioritized calls, respectively. Here, we reserve $C - T$ channels exclusively for handoff calls of handoff-prioritized service class. Therefore, the number of handoff-guaranteed service class is subject to the upper bound of T. The state space can be written as,

$$\Lambda = \{\mathbf{x}, |0 \leq x_G + x_P \leq C \text{ and } 0 \leq x_G \leq T\}.$$

The handoff-guaranteed service can be divided into N sub-dimensions, where the mth sub-dimension has an m-stage Erlangian departure $(0 \leq m \leq N)$, denoting the handoff-guaranteed call that is going to enter this cell as its mth cell. The state of the mth sub-dimension x_{Gm} denotes the number of calls in this sub-dimension. Then, $\sum_{m=1}^{N} x_{Gm} = x_G$.

Assuming that the arrival and the departure of each dimension are independent of each other, the steady state distribution $P(\mathbf{x})$ can be obtained numerically. Given the steady state distribution $P(\mathbf{x})$ and threshold T for handoff-prioritized calls, we can derive the new call blocking probability and the handoff dropping probability for both types of calls as follows.

For the handoff-prioritized call, the new call blocking probability P_{BP} for a given cell is the probability that T or more channels of this cell are occupied, i.e.

$$P_{BP} = \sum_{\mathbf{x} \in \Lambda, x_G + x_P \geq T} P(\mathbf{x}). \tag{4.6}$$

The handoff dropping probability P_{DP} is the probability that all C channels are being occupied, i.e.

$$P_{DP} = \sum_{\mathbf{x} \in \Lambda, x_G + x_P = C} P(\mathbf{x}). \tag{4.7}$$

For the handoff-guaranteed call, the new call blocking probability $P_{BG}(n)$ is the probability that at least one of the n cells it is going to enter has no available channel for the new call request. Under the assumption of a unified cell load in all cells, we have

$$P_{BG}(n) = 1 - (1 - P_{BP})^n, \tag{4.8}$$

where P_{BP} is calculated in (4.6). Here, the handoff dropping probability $P_{DG} = 0$.

4.2.3   System with Three Service Classes

Let us consider a cell with C channels offering all three connection-level service classes described above. The system can be analyzed by using a multi-dimensional model as follows. The state of a system is the vector $\mathbf{x} = \{x_G, x_P, x_U\}$, where x_G, x_P and x_U are the numbers of occupied (being used or reserved) channels by handoff-guaranteed, handoff-prioritized and handoff-undeclared calls, respectively. Here, we reserve $C - T$ channels exclusively for handoffs of handoff-prioritized service class. Therefore, the total number of handoff-guaranteed and handoff-undeclared calls is subject to the upper bound of T. The state space can be written as,

$$\Lambda = \{\mathbf{x}, |0 \leq x_G + x_P + x_U \leq C \text{ and } 0 \leq x_G + x_U \leq T\}.$$

The handoff-guaranteed service can be divided into N sub-dimensions, where the mth sub-dimension has an m-stage Erlangian departure ($0 \leq m \leq N$), denoting the handoff-guaranteed call that is going to enter this cell as its mth cell. The state of the mth sub-dimension x_{Gm} denotes the number of calls in this sub-dimension. Then, $\sum_{m=1}^{N} x_{Gm} = x_G$.

Assuming that the arrival and the departure of each dimension are independent of each other, the steady state distribution $P(\mathbf{x})$ can be obtained numerically. Given the steady state distribution $P(\mathbf{x})$ and threshold T for handoff-prioritized calls, we can derive the new call blocking probability and the handoff dropping probability for each type of calls as follows.

For the handoff-undeclared call, both the new call blocking probability P_{BU} and the handoff dropping probability P_{DU} are the same, i.e.

$$P_{BU} = P_{DU} = \sum_{x_G + x_P + x_U \geq T} P(\mathbf{x}), \qquad (4.9)$$

which is the probability of the cell in states with equal or more than T channels occupied.

For the handoff-prioritized call, the new call blocking probability P_{BP} for a given cell is the the same as that of handoff-undeclared calls, i.e., the probability that T or more channels of this cell are occupied.

$$P_{BP} = \sum_{\mathbf{x} \in \Lambda, x_G + x_P + x_U \geq T} P(\mathbf{x}). \qquad (4.10)$$

The handoff dropping probability P_{DP} is the probability that all C channels are being occupied, i.e.

$$P_{DP} = \sum_{\mathbf{x} \in \Lambda, x_G + x_P + x_U = C} P(\mathbf{x}). \qquad (4.11)$$

For the handoff-guaranteed call, the new call blocking probability $P_{BG}(n)$ is the probability that at least one of the n cells it is going to enter has no available channel for the new call request. Under the assumption of a unified cell load in all cells, we have

$$P_{BG}(n) = 1 - (1 - P_{BU})^n, \qquad (4.12)$$

where P_{BU} is calculated in (4.9). On the other hand, the handoff is guaranteed not being dropped for this type of calls by isolated reservation, i.e., the handoff dropping probability P_{DG} is

$$P_{DG} = 0. \tag{4.13}$$

The above multi-dimensional M/Er/C/C queuing model has been used to explain our system, especially for differentiating the handoff dropping probability requirements of different service classes. In fact, we are not aware of any generalized closed form solution for this. However, the numerical solution can be obtained by existing queuing model softwares, such as QTS [10]. Alternatively, the numerical solution can be obtained by discrete-event simulation, as we have done in the following section 4.3, using the OPNET simulator.

4.3 Simulation and Discussion

To evaluate the performance of the proposed adaptive resource management system, a network model of a single cell with channel capacity C was constructed in the OPtimized Network Engineering Tool (OPNET), which is a discrete-event driven simulator. The simulation conducted here consists of the following three service classes: handoff-guaranteed CBR, handoff-prioritized CBR, and non-real-time UBR best-effort services. Based on the modeling analysis conducted in Section 4.2, a number of call generators generated Poisson arrivals of new and handoff call requests from different service classes. We use the following notations for the simulation parameters throughout this section.

- For the handoff-guaranteed CBR service:

  - N: the maximum number of cells a handoff-guaranteed call traverses in its lifetime.
  - λ_{Gi}: the mean arrival rates of new call requests from handoff-guaranteed calls entering this cell as its ith ($1 \leq i \leq N$) cell.
  - η_G: the mean of the exponentially distributed cell residence time of the handoff-guaranteed calls. Thus, the

corresponding channel holding time of handoff-guaranteed calls have i-stage Erlangian distributions, $(1 \leq i \leq N)$, and the mean channel holding times are $i * \eta_G$.

- For the handoff-prioritized CBR service:

  - λ_{Pn}: the mean arrival rate of new call requests from handoff-prioritized service class.

  - λ_{Ph}: the mean arrival rate of handoff requests from handoff-prioritized service class.

  - η_P: the mean of the exponentially distributed cell residence times of the handoff-prioritized calls.

- For the non-real-time UBR best-effort service:

  - λ_B: the mean packet arrival (Poisson) rate.

  - S: the packet size in bytes per packet.

4.3.1   A Typical Scenario

To simulate a typical scenario, system parameters were set as follows. The capacity of each cell is $C = 64$ channels with data rate 4800 bps per channel. For the handoff-guaranteed service, $N = 6$, $\lambda_{G1} = \lambda_{G2} = 0.05$ call/sec, $\lambda_{G3} = \lambda_{G4} = 0.02$ call/sec, and $\lambda_{G5} = \lambda_{G6} = 0.01$ call/sec, $\eta_G = 25$ sec. For the handoff-prioritized service, $\lambda_{Pn} = 0.2$ call/sec, $\lambda_{Ph} = 0.1$ call/sec, $\eta_P = 50$ sec. The target handoff dropping probability P_{DP_tar} of the handoff-prioritized service was set to 0.01. For the UBR best-effort service, $\lambda_B = 30$ packets/sec with fix-sized packets of $S = 1K$ Bytes/packet.

We simulated three types of real-time multimedia traffic, i.e. voice, audio and video, each requiring 1, 2 and 4 channels from the network, respectively. Among the generated handoff-guaranteed and handoff-prioritized calls, we randomly select 50% as voice, 25% as audio and the remaining 25% as video applications. The performance in terms of the new call blocking probability, the handoff dropping probability and channel utilization were compared for adaptive and non-adaptive applications. In the experiments for adaptive applications, audio and video calls were assumed to reduce their rate at 1 channel under congestion. The results for a 20-hour simulation are shown in Figs. 4.2-4.5. We use HG and HP to denote the handoff-guaranteed and the handoff-prioritized service classes, respectively.

A measurement-based dynamic guard channel scheme is used for the HP service class to achieve the target handoff dropping probability after a short initial unstable stage as shown in Fig. 4.2 (a) for both adaptive and non-adaptive applications. The details of the scheme will be discussed in Chapter 5. The dynamic change of the threshold for achieving the target handoff dropping probability is shown in Fig. 4.2 (b). The value of the threshold fluctuates more for non-adaptive applications, since rate-adaptive applications have the ability to adapt to network conditions.

There is no handoff dropping for HG service class due to band-width reservation. Figs. 4.3 (a) and (b) show the new call blocking probability for HG and HP service classes, respectively. The HG service has a higher new call blocking probability than HP service, since it requires more resources to be reserved to ensure the zero handoff dropping probability. The comparison between adaptive and non-adaptive applications in Fig. 4.3 shows that the new call blocking of both real-time service classes are reduced significantly by exploiting rate adaptability of multimedia applications. Similarly, the overall handoff dropping probability (Fig. 4.4 (a)) and new call blocking probability (Fig. 4.4 (b)) of real-time services are lower for adaptive applications.

Fig. 4.5 shows the channel utilization of real-time and non real-time UBR service classes. Due to resource reservation, real-time service classes cannot fully utilize channels to maintain the desired connection-level QoS, even taking rate-adaptive applications into account. However, the UBR best-effort service class can exploit the reserved but unused resources as shown in Fig. 4.5 (b), thus improving the overall utilization for both adaptive and non adaptive cases. The channel utilization of real-time service classes (Fig. 4.5 (a)) increases slightly when incorporating adaptive applications. Fig. 4.6 illustrates the percentage of reduced bandwidth real-time calls.

4.3.2   Varying Traffic Loads

To investigate the system performance under different traffic loads, we conducted two sets of simulations, one with varying real-time traffic load and the other with varying non real-time traffic load as shown in Figs. 4.7 and 4.8, respectively.

In the first case, the new call arrival rate λ_{Pn} of the handoff-prioritized service was set to vary from 0.05 to 0.95 calls/sec, while maintaining all other settings as those in the above typical scenario. The results are shown in Fig. 4.7. Fig. 4.7 (a) shows that the handoff dropping probabilities of HP and HG service classes remain at the target with increasing real-time traffic. Fig. 4.7 (b) shows that the new call blocking probability of both real-time service classes increases with increasing real-time traffic. Fig. 4.7 (c) shows that the bandwidth utilized by real-time service classes increases with the increase in the real-time traffic, while the throughput of non real-time UBR traffic decreases. As a result, channel utilization in our scheme is quite high for real-time traffic, even without the best-effort service. This is achieved by our dynamic guard channel scheme and the rate adaptation. Also the decrease in throughput of the best-effort class traffic shows that the real-time HG and HP traffic effectively preempts the on-going best-effort traffic. Since the total bandwidth utilization is 100%, it is clear that the non real-time traffic is able to fully exploit the resource not being used by real-time traffic.

In the second case, the packet arrival rate λ_B of the best-effort UBR service was set to vary from 5 to 100 packets/sec, while maintaining all other parameters the same as the above typical scenario. The simulation results are shown in Fig. 4.8. We can see from this figure that the traffic load of the non real-time UBR service class does not influence connection-level QoS parameters of real-time service classes, but influences the throughput of non real-time UBR traffic itself.

The results shown in Figs. 4.7 and 4.8 demonstrate that the system gives preference to real-time services to achieve their required QoS. On the other hand, the non real-time service improves the total utilization of network resources. Here we assume that a mechanism exists in MAC layer to permit the best-effort class packet occupy the unused channels, preempt them by real-time traffic and take care of collisions.

4.4 Conclusion and Future Work

In this chapter, we discussed the adaptive resource management mechanisms to provide QoS for wireless multimedia networks. We proposed a comprehensive service model for wireless multimedia applications. The service model takes into account a primary connection-level QoS measurement, the handoff dropping probability, and the traffic characteristics of different wireless multimedia applications. Based on the proposed service model, adaptive resource allocation is performed by applying different call admission control and resource reservation schemes to each service class appropriately.

The system is adaptive to the network traffic conditions as well as the nature of multimedia applications. Some detailed design issues, such as policing and adaptation cost, are addressed for the implementation of the proposed system. It is demonstrated by simulations that the proposed service model can accommodate more realistic services and achieve better network utilization. By integrating the adaptation of multimedia applications, the scheme also improves total user service satisfaction in a mobile wireless environment with varying and limited bandwidth resources. Thus, multimedia traffic can get better QoS guarantees at different levels, while achieving high network utilization.

It is worth mentioning that the proposed system is flexible in its detailed implementation. Different call admission control, resource reservation schemes, policing mechanisms, and adaptation algorithms can be designed and incorporated under this generic framework. In our specific implementation, we considered the straightforward schemes to simplify the network complexity. However, more advanced and sophisticated mechanisms, including complete connection-level and application-level service models, could further improve the overall system performance

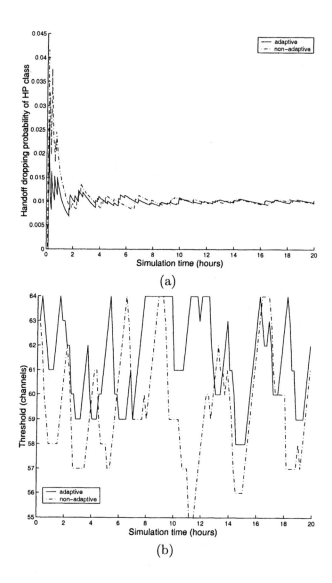

Figure 4.2: Comparison for adaptive and non-adaptive applications under a typical scenario: (a) the handoff dropping probability of the HP class, (b) the threshold.

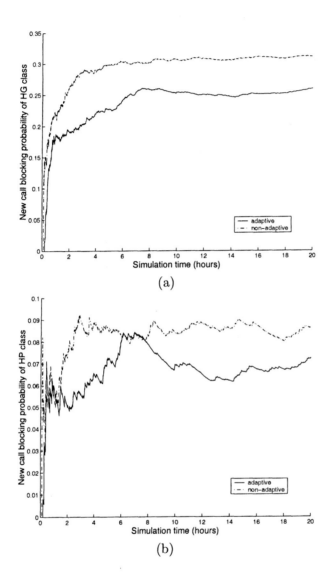

Figure 4.3: Performances comparison for adaptive and non-adaptive applications under a typical scenario:(a) the new call blocking probability of the HG class, (b) the new call blocking probability of the HP class.

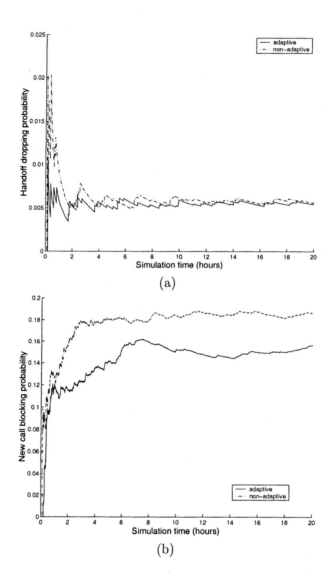

Figure 4.4: Performances of real-time services comparison for adaptive and non-adaptive applications under a typical scenario:(a) the handoff dropping probability, (b) the new call blocking probability.

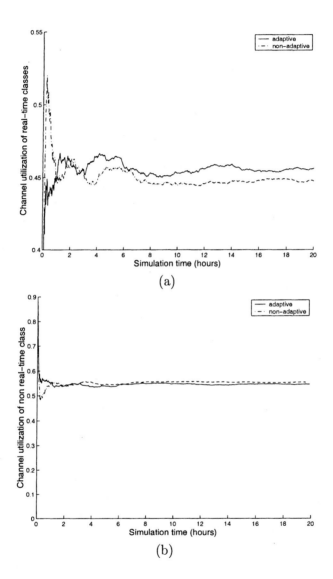

Figure 4.5: Channel utilization under a typical scenario:(a) real-time
service classes, (b) non real-time service class.

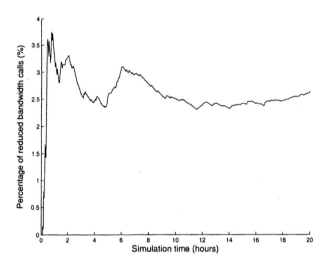

Figure 4.6: The percentage of reduced-bandwidth real-time calls.

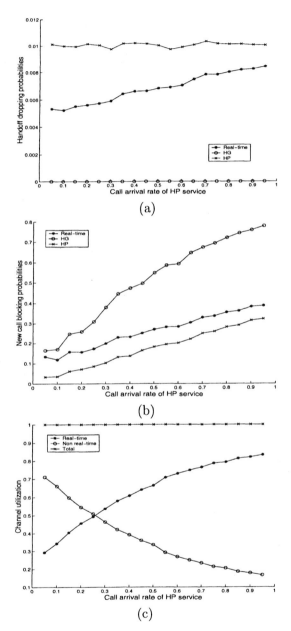

Figure 4.7: Performances under varying real-time traffic load:(a) the handoff dropping probabilities, (b) the new call blocking probabilities, and (c) channel utilization.

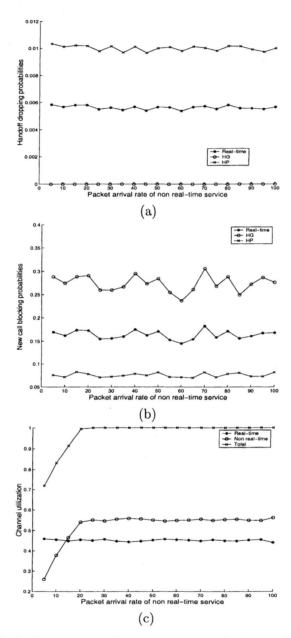

Figure 4.8: Performances under varying non real-time traffic load:(a) the handoff dropping probabilities, (b) the new call blocking probabilities, and (c) channel utilization.

BIBLIOGRAPHY

[1] S. Shenker, D. D. Clark, and L. Zhang, "Services or infrastructure: why we need a network service model," In *Proc. of the 1st International Workshop on Community Networking: Integrated Multimedia Services to the Home*, pages 145–149, 1994.

[2] R. Handel, S. Schroder, and M. Huber, *ATM Networks: Concepts, Protocols, Applications*. Addison Wesley Longman Ltd., Mar. 1998.

[3] R. Braden, D. Clark, and S. Shenker, "Integrated service in the internet architecture: an overview," IETF RFC1633, 1994.

[4] K. Nichols, V. Jacobson, and L. Zhang, "A two-bit differentiated services architecture for the internet," Internet Draft ⟨draft-nichols-diff-svr-arch-00.txt⟩, Nov. 1997.

[5] C. Oliveira, J. B. Kim, and T. Suda, "An adaptive bandwidth reservation scheme for high-speed multimedia wireless networks," *IEEE Journal on Selected Areas in Communication*, 16(6):858–874, Aug. 1998.

[6] A. K. Talukdar, B. R. Badrinath, and A. Acharya, "Integrated services packet networks with mobile hosts: Architecture and performance," *Wireless Networks*, 5(2):111–124, Mar. 1999.

[7] ISO/IEC JTC1/SC29/WG11, "Generic coding of moving pictures and associated audio information," ISO/IEC International Standard 13818, Nov. 1994.

[8] ISO/IEC JTC1/SC29/WG1, "JPEG 2000 part 1 final committee draft version 1.0," ISO/IEC International Standard N1646R, Mar. 2000.

[9] ISO/IEC JTC1/SC29/WG11, "Overview of the MPEG-4 standard," ISO/IEC N3747, Oct. 2000.

[10] D. Gross and C. M. Harris, *Fundamentals of Queueing Theory*, John Wiley & Sons, 3rd Edition, 1998.

Chapter 5

DYNAMIC CALL ADMISSION CONTROL SCHEMES

Call admission control (CAC) is the mechanism for making decision about accepting a user's connection request according to certain criteria. Generally, these criteria are based on available and requested resources to ensure the demanded service quality. As described in Chapter 4, appropriate call admission control and corresponding resource reservation schemes are crucial for QoS provisioning in wireless networks. Many call admission control and resource allocation schemes have been proposed for wireless/mobile networks. Optimal stationary call admission control schemes have been investigated for single-service [1] and multi-service [2] wireless networks. However, stationary schemes have several disadvantages. First, they are less efficient in real networks where the traffic pattern is changing from time to time. Second, the computational complexity involved in the optimization techniques is usually too high to be practical. Third, stationary solutions generally cannot achieve multiple constraints at the same time without sacrificing resource utilization to meet the most strict ones. To overcome these shortcomings, we focus on dynamic CAC and resource allocation schemes to provide a variety of QoS guarantees for wireless multimedia services.

As measurement-based approaches, our proposed dynamic CAC and resource allocation schemes have the following features. First, they are robust to the changing network conditions. They do not make any assumptions about the traffic distribution pattern. Second, the schemes are computationally efficient in a distributed manner since the base station in each cell can monitor its own performance and take appropriate action accordingly. Third, they can be applied

to multimedia applications with heterogenous bandwidth requirements. Fourth, compared with other measurement-based schemes
that rely on the measurement of current traffic conditions [3, 4], our
proposed dynamic schemes are based on measurements of the current QoS metrics directly, which gives more accurate information
and enables more efficient control to achieve the design objectives.
At last, there is no need for either any information exchange among
neighboring cells, as done in [3], or accurate location-tracking information, as required in [4]. Thus, our proposed scheme are practically
applicable.

In the remaining of this chapter, the proposed dynamic CAC
schemes for CBR and VBR traffic are discussed in details. Simulations are conducted in OPNET simulator to show the performance
of the proposed schemes.

5.1   Dynamic CAC Schemes for CBR Traffic

Constant Bit Rate (CBR) traffic requires a certain amount of bandwidth resource available during the entire lifetime of the connection.
From this point of view, it is similar to the connection-oriented circuit switching service provided in traditional wireless communication
networks. As long as the required resource is available, the admitted CBR traffic will have packet-level QoS guarantees in terms of
delay, delay jitter and packet loss. Therefore, the connection-level
QoS for CBR traffic provides the basis for packet-level QoS, thus
is especially important for CBR traffic. In this section, we address
the provisioning of connection-level QoS, measured by the new call
blocking probability P_B and the handoff dropping probability P_D,
for CBR traffic in wireless multimedia networks.

5.1.1   Problem Formulation

A dynamic CAC scheme with connection-level QoS guarantees could
have different objectives and constraints. In particular, we consider
the following two:

A minimize P_B subject to the P_D constraint $P_D \leq P_D^{tar}$,

B minimize P_D subject to the P_B constraint $P_B \leq P_B^{tar}$,

where P_D^{tar} (P_B^{tar}) are the predefined target handoff dropping probability (new call blocking probability).

The first one has been found in many existing CAC schemes proposed so far [5, 6, 7, 3, 4]. It guarantees a certain level of service to already admitted users by placing a hard constraint on the P_D and tries to maximize network utilization by minimizing the P_B. To network service providers, this means maximizing the net revenue. Therefore, it is of great practical interest in the real world. For this reason, we first develop our dynamic CAC scheme based on this objective. Then, we extend our analysis and proposed algorithm to solve the second problem, which may be of less practical significance.

5.1.2 Proposed Dynamic CAC Schemes

Our proposed dynamic CAC scheme is based on the priority handoff scheme, also called the guard channel scheme, introduced in [8]. It has been proved to be the optimal stationary CAC policy for minimizing a linear cost function. Moreover, the fractional guard channel scheme , as one of its variant with randomization, has been proved to be the optimal stationary CAC policy for the above objective A under certain conditions [1]. In fact, our proposed dynamic guard channel scheme could also be viewed as its another variant with randomization over time. It can be adaptive to the non-stationary varying traffic condition.

Considering a single cell with a fixed amount of bandwidth capacity of C channels, the traditional guard channel scheme gives a higher priority to the handoff call request as compared to the new call request by reserving a portion of the channel resource for handoff calls. More specifically, a new call request is admitted only when there are less than T channels occupied, where T is a threshold between 0 and C. On the other hand, a handoff request is rejected only when all C channels are occupied. As a result, $C - T$ channels are the *guard channels*, which are used only by handoff calls.

The threshold T for the guard channel scheme plays an important role in the P_B and P_D performance. Increasing T results in less reserved resources for handoff calls, and increases the P_D. At the same time, the P_B decreases accordingly because more resources become available for new arriving calls. On the other hand, decreasing T has the opposite effect. Moreover, the overall system utilization could also be influenced by the value of T. Reserving more resources than needed for handoff calls results in lower system utilization since reserved resources cannot be used by new call requests. Apparently, the selection of T, i.e. the reserved resources for handoff calls, should dynamically vary with changing traffic conditions.

The P_B and the P_D of the guard channel scheme have the following properties, which were proved in [9].

Property 1: The P_D is strictly increasing with threshold T increasing, i.e.

$$If \quad T1 < T2, \quad then \quad P_D(T1) < P_D(T2) \qquad (5.1)$$

Property 2: The P_B is strictly decreasing with threshold T increasing, i.e.

$$If \quad T1 < T2, \quad then \quad P_B(T1) > P_B(T2) \qquad (5.2)$$

With the above properties, it is easy to derive the following corollary. The P_B is strictly decreasing with the P_D increasing, caused by changing of threshold T. The corollary implies that minimization of the P_B under the constraint of $P_D \leq P_D^{tar}$ could be achieved by keeping P_D as close to P_D^{tar} as possible.

Minimize P_B subject to P_D Constraint

Thus, the objective of the dynamic guard channel scheme can be stated as follow. For a given system with a target handoff dropping probability P_D^{tar}, threshold T should be selected to keep the resulting P_D as close to P_D^{tar} as possible without exceeding it while the P_B should be minimized. In the following sections, we propose a measurement-based algorithm that aims at achieving the above objective by dynamically adjusting T.

Basically, the proposed dynamic CAC scheme is based on measurements of the current P_D. The proposed scheme for minimizing P_B subject to P_D constraints works as follows. At the beginning, an initial value of T_{init} is selected for a given cell. The base station of this cell monitors its P_D. When the measured P_D reaches or exceeds the target value P_D^{tar}, T is decreased by one unit so that more guard channels are reserved for handoff requests. Otherwise, we increase T by one unit to admit more new call requests. The proposed algorithm for this objective is summarized with the pseudo-code given in Fig. 5.1, where $B_{occupied}$ denotes the currently occupied bandwidth, and B_{req} denotes the required bandwidth by the call request. Some design issues are discussed in the following.

Prompt Decreasing and Timer Increasing (PDTI)   An important design issue that affects the performance of the scheme is how often and when to update the measurement of P_D to dynamically adjust T. A frequent update can keep pace with changing traffic conditions. If the T is not updated fast enough to match the variation in system conditions, it could result in lower channel utilization due to late increase in T, or higher P_D due to late decrease in T. However, a new value of T does not immediately affect the measured P_D. Thus, very frequent T updates could cause unnecessary fluctuations and burden to the system. Therefore, a tradeoff is desired between fast response and system stability.

We use a strategy called *prompt-decreasing/timer-increasing* (PDTI) to choose the timing for updating the value of T in the proposed scheme. Whenever a handoff is dropped, the base station checks the current value of P_D and decreases T if $P_D \geq P_D^{tar}$. Since only handoff dropping events could drive the P_D higher than the target value, this 'prompt-decreasing' strategy can immediately take the corrective action by decreasing T to avoid further handoff droppings. At the same time, a timer is set. If there is no further handoff droppings, upon the expiration of the timer, the base station checks if $P_D < P_D^{tar}$. If it is true, the threshold T is increased to improve channel utilization. Otherwise, it renews the timer. Before the timer's expiration, if there are further handoff droppings, the base station again checks P_D. It further decreases T and resets the timer if necessary.

Timer Setting   The timer setting allows the effect of decreasing T being reflected in the change of P_D, since it may take several successful handoffs before measured P_D falls below the P_D^{tar}, or before it gets

01   INITIALIZE $T = C$
02   SET TIMER
03   WAIT FOR CALL REQUEST ARRIVAL
04   **If** NEW CALL REQUEST ARRIVES
05   **If** $(B_{occupied} + B_{req}) \leq (C - T)$
06   ADMIT NEW CALL WITH RATE B_{req}
07   **Else**
08   REJECT NEW CALL REQUEST
09   **If** HANDOFF CALL REQUEST ARRIVES
10   **If** $(B_{occupied} + B_{req}) \leq C$
11   ADMIT HANDOFF CALL WITH RATE B_{req}
12   **Else**
13   REJECT HANDOFF REQUEST
14   UPDATE P_D
15   **If** $P_D \geq P_D^{tar}$ AND $T > 0$
16   DECREASE T
17   RESET TIMER AND GO BACK TO 03
18   **If** TIMER EXPIRES
19   INCREASE T
20   RESET TIMER AND GO BACK TO 03
21   **Else** GO BACK TO 03

Figure 5.1: The proposed dynamic guard channel call admission control algorithm for minimizing P_B with P_D constraint.

to the steady state. The timer is renewed when the P_D continues to drop down, which is indicated by no handoff dropping, until it falls below P_D^{tar}, which triggers an increase in T, or further handoff dropping happens when $P_D \geq P_D^{tar}$, which means reserved resources are not enough for handoff and further decrease in T is needed. Thus, it avoids the unnecessary fluctuation of T and provides necessary updating. The value of the timer is the maximum time that the system could be under-utilized, before it starts to increase T to improve its utilization when the P_D is lower than the target.

Initial Threshold Another issue is the selection of the initial threshold T. Theoretically, it can be derived from mathematical analysis conducted in the previous section according to given P_D^{tar}. This approach can lead the system into a steady state quickly. However, it requires modeling parameters such as the new call and the handoff arrival rates, the cell residence time, etc. Practically, when these parameters are not available, we can simply begin with a non-priority handoff scheme, where T_{init} is set to the capacity of the cell C. By applying the proposed dynamic scheme, it will reach a steady state after some time. Starting with $T_{init} = C$ maximizes system utilization from the beginning. However, it might be achieved at the cost of violating the P_D^{tar} bound in the initial period.

Achievable Constraints The achievable range of the P_D constraint depends on the system configuration (*e.g.* the channel capacity) and the traffic pattern (*e.g.* new/handoff call arrival/departure rates). There exists some range that P_D^{tar} cannot be satisfied if the traffic is too heavy under a given system capacity. The lowest achievable P_D^{tar} can be found by setting $T = 0$, which means all C channels are exclusively used for handoff call requests. Under the assumptions given in Section 3.1, this lower bound can be found as

$$P_D \geq P(C)_{T=0} = \frac{\frac{(\lambda_h/\eta)^C}{C!}}{\sum_{i=0}^{C} \frac{(\lambda_h/\eta)^i}{i!}}. \tag{5.3}$$

For the case with multiple services, the achievable range of P_D^{tar} is still lower bounded by $P(C)_{T=0}$. Generally speaking, a wireless network is designed such that the P_D^{tar} is achievable under its traffic condition.

Minimize P_D Subject to P_B Constraint

Let us consider another optimization problem by swapping the role of P_B and P_D in the objective function. The corresponding dynamic guard channel scheme is to determine threshold T dynamically so that the resulting P_B is kept as close to the target level P_B^{tar} as possible without exceeding it. The dynamic guard channel algorithm proposed earlier can be modified to solve this problem as shown in Fig. 5.2.

In this algorithm, the P_B is measured by the base station upon every new call blocking event such that threshold T can be increased promptly to ensure $P_B \leq P_B^{tar}$, while the timer is set to minimize the P_D by decreasing T whenever T is unnecessarily high for achieving P_B^{tar}.

The achievable range of the P_B constraint again depends on the system configuration (*e.g.* the channel capacity) and the traffic pattern (*e.g.* new/handoff call arrival/departure rates). There exists some range that P_B^{tar} cannot be met if the traffic is too heavy for the given system capacity. The lowest achievable P_B^{tar} can be found by setting $T = C$, which means all the C channels are available for new call requests. Under the same assumptions given in Section 3.1, the value of achievable P_B^{tar} is bounded by $P(C)$ with $T = C$. That is,

$$P_B \geq P(C)_{T=C} = \frac{\frac{((\lambda_n+\lambda_h)/\eta)^C}{C!}}{\sum\limits_{i=0}^{C} \frac{((\lambda_n+\lambda_h)/\eta)^i}{i!}}. \qquad (5.4)$$

When considering multiple services, the achievable range of P_B^{tar} is still bounded by $P(C)$ with $T = C$. Generally, a wireless network is designed such that the target P_B are achieved for its traffic condition. An algorithm to find the minimum cell capacity (in terms of the number of channels) for traditional single-service wireless networks under a certain traffic model was proposed in [1]. Considering varying traffic and multi-service characteristics in a wireless multimedia network, we discuss below an algorithm for dynamic channel

01 INITIALIZE $T = C$
02 SET TIMER
03 WAIT FOR CALL REQUEST ARRIVAL
04 **If** NEW CALL REQUEST ARRIVES
05   **If** $(B_{occupied} + B_{req}) \leq (C - T)$
06   ADMIT NEW CALL WITH RATE B_{req}
07   **Else**
08   REJECT NEW CALL REQUEST
09   UPDATE P_B
10   **If** $P_B \geq P_B^{tar}$ AND $T < C$
11   INCREASE T
12   RESET TIMER AND GO BACK TO 03
13 **If** HANDOFF CALL REQUEST ARRIVES
14   **If** $(B_{occupied} + B_{req}) \leq C$
15   ADMIT HANDOFF CALL WITH RATE B_{req}
16   **Else**
17   REJECT HANDOFF REQUEST
18 **If** TIMER EXPIRES
19   DECREASE T
20   RESET TIMER AND GO BACK TO 03
21 **Else** GO BACK TO 03

Figure 5.2: The proposed CAC algorithm with a dynamic guard channel to minimize P_D under the P_B constraint.

assignment that meets both P_B and P_D constraints in the following section.

5.1.3   Dynamic Channel Assignment Subject to Joint P_B and P_D Constraints

Dynamic channel assignment (DCA) [10] among adjacent cells in a wireless communication system achieves more efficient use of system resources at a cost of higher complexity and more state information of neighboring cells. Various combinations of permanent channel assignment, channel borrowing, shared pools of channels, channel ordering, channel reassignment, and dynamic adjustment of parameters have been suggested to realize DCA [11][12][13]. In this section, we address the basic question in DCA: from a single cell's point of view, how many channels should be assigned to the given cell in order to achieve the target P_B and P_D constraints? This can be formulated as follows.

Minimize cell capacity C subject to the following constraints

$$P_B(\phi) \leq P_B^{tar} \text{ and } P_D(\phi) \leq P_D^{tar}$$

where ϕ is the corresponding CAC policy under C.

Following the similar idea presented in the previous section, a new dynamic algorithm is proposed to solve the above question of DCA as illustrated in 5.3.

In this algorithm, both capacity C and threshold T of a cell are adjusted dynamically such that the joint constraints $P_B \leq P_B^{tar}$ and $P_D \leq P_D^{tar}$ can be satisfied, while keeping C as small as possible. The base station measures both the P_B and the P_D upon new call blocking and handoff dropping events. There are three cases when the value of C should be increased. First, if the measured $P_B > P_B^{tar}$ when $T = C$, it is impossible to increase T further without increasing C to get lower P_B. Second, if the measured P_B and P_D both exceed the targets, then either increasing or decreasing T alone can not make both P_B and P_D lower than their target values. Third, if the measured $P_D > P_D^{tar}$ when $T = 0$, then it is impossible to decrease T without increasing C to get lower P_D. In a practical wireless communication network, where the new call traffic is generally heavier than the handoff traffic, the last case is unlikely to happen.

---

01 INITIALIZE $T = C$
02 SET TIMER
03 WAIT FOR CALL REQUEST ARRIVAL
04 **If** NEW CALL REQUEST ARRIVES
05   **If** $(B_{occupied} + B_{req}) \leq (C - T)$
06   ADMIT NEW CALL WITH RATE B_{req}
07   **Else**
08   REJECT NEW CALL REQUEST
09   UPDATE P_B
10   **If** $P_B \geq P_B^{tar}$
11            $T = T + 1$
12   **If** $T > C$
13                $C = C + 1$
14   RESET TIMER AND GO BACK TO 03
15 **If** HANDOFF CALL REQUEST ARRIVES
16   **If** $(B_{occupied} + B_{req}) \leq C$
17   ADMIT HANDOFF CALL WITH RATE B_{req}
18   **Else**
19   REJECT HANDOFF REQUEST
20   UPDATE P_D
21   **If** $P_D \geq P_D^{tar}$
22   **If** $P_B \geq P_B^{tar}$ OR $T = 0$
23                $C = C + 1$
24   **Else**
25                $T = T - 1$
26   RESET TIMER AND GO BACK TO 03
27 **If** TIMER EXPIRES
28    $C = C - 1$
29   **If** $T > C$
30        $T = C$
31   RESET TIMER AND GO BACK TO 03
32 **Else** GO BACK TO 03

Figure 5.3: The proposed DCA algorithm with both P_B and P_D constraints.

The timer setting detects when C is higher than necessary to meet the targets of P_B and P_D. Upon the timer expiration, if both P_B and P_D have been found below their targets for the timer's period, we decrease C to the minimum necessary value.

With C that satisfies both P_B and P_D constraints under the current traffic condition, the dynamic change of T between 0 and C ensures $P_B \leq P_B^{tar}$ and $P_D \leq P_D^{tar}$. If the base station detects that $P_B > P_B^{tar}$, it increases T to lower P_B at the cost of higher P_D within its constraints. Otherwise, if the base station detects that $P_D > P_D^{tar}$, it decreases T to lower P_D at the cost of higher P_B within its constraint.

5.2 Dynamic CAC Schemes for VBR Traffic

In the above section, we developed the dynamic call admission control schemes for the connection-level QoS guarantee for CBR traffic. The schemes are also applicable to traditional wireless communication networks using the circuit-switching technology for voice communications, where the connection-level QoS is the primary concern. In this section, we investigate the call admission control for VBR traffic in cellular networks considering both connection-level and packet-level QoS.

Packet-level QoS is usually measured by a set of parameters such as delay/delay jitter, error/loss and throughput, etc. The wireless network of our interest in this work is the last-hop type, where the infrastructure consists of a wireless access network and a wired backbone network such as the Internet. Although the wireless access network may result in longer delay in general, the major part of delay jitter comes from the multi-hop wired Internet. QoS in this aspect has been investigated extensively in the context of IP networks. Extension to the last hop of wireless access has also been studied [14, 15]. For this reason, we consider the packet loss probability as the primary packet-level QoS metric in this section.

It is well known that a small amount of packet loss is tolerable in most VBR multimedia applications. Most multimedia applications with the VBR source such as video, audio and voice communications could tolerate a certain degree of packet loss with little perceivable quality degradation. Generally speaking, the smaller the packet loss probability is, the higher quality could be perceived by the end user. In this context, a trade-off between packet-level and connection-level QoS and their joint optimization is important for the overall QoS provisioning.

In a wireless network, packet loss occurs at different layers. In the physical layer, the error-prone radio channel causes data corruption. Advanced channel coding schemes have been developed to solve this problem [16, 17, 18]. In the link layer, competition for access to the shared radio resource results in data collision. Multiple access control (MAC) schemes with QoS considerations have been developed to avoid this problem [19, 20]. In the network layer, packet loss is primarily caused by congestions due to scare network resources. Our work addresses this problem and develops a network layer mechanism to jointly optimize packet-level QoS (in terms of congestion-caused packet loss probability) and connection-level QoS (in terms of the P_B and P_D).

5.2.1 VBR Traffic Model

For mathematical tractability, each VBR traffic flow is modeled by a two-state on-off model . It is assumed in this model that the source traffic is in either the on state or the off state. In the on state, data traffic is generated at a peak rate, while in the off state, no traffic is generated. The probability of the source in the on state is denoted by P_{on}. Furthermore, we assume that all VBR connections have the unit peak rate, and are independent of each other. Mathematically, the ith VBR traffic is a random process $X_i(t)$, which are independent identically distributed random variables for any set of i and t with $P\{X_i(t) = 1\} = P_{on}$ and $P\{X_i(t) = 0\} = 1 - P_{on} = P_{off}$.

The current load $L(t)$, defined as the total traffic in the "on" state, is then a random process obtained by summation of all the active connections, $i.e.$,

$$L(t) = \sum_{i=i}^{N} X_i(t), \tag{5.5}$$

where N is the total number of VBR connections being carried in the system. For any t, $L(t)$ has a binomial distribution $B(N, P_{on})$, i.e.

$$P\{L(t) = k\} = \binom{N}{k} P_{on}^k (1 - P_{on})^{N-k}. \qquad (5.6)$$

Suppose that the system capacity is C. Then, the average load loss is the expectation of $L(t)$ exceeding C, which can be written as

$$E(L(t) - C) = \sum_{m=1}^{N-C} m P\{L(t) = C + m\}. \qquad (5.7)$$

The average total load is

$$E(L(t)) = \sum_{k=1}^{N} k P\{L(t) = k\} = N P_{on}. \qquad (5.8)$$

Thus, the load loss rate $R_L(N)$ with N calls in the system is

$$R_L(N) = \frac{E(L(t) - C)}{N P_{on}}. \qquad (5.9)$$

Assume that the new and the handoff connection arrivals are Poisson distributed with rates λ_n and λ_h, respectively. The cell residence time of both connections is exponentially distributed with mean $\frac{1}{\eta}$. Let $\lambda = \lambda_n + \lambda_h$, and $\lambda_n = \alpha\lambda$. The corresponding Markov chain gives the stationary distribution as

$$\pi(N) \propto \frac{(\lambda/\eta)^N}{N!} \prod_{i=1}^{N} (\alpha P_{an}(i - 1) + (1 - \alpha) P_{ah}(i - 1)), \qquad (5.10)$$

such that

$$\sum_{i=0}^{N} \pi(i) = 1, \qquad (5.11)$$

where $P_{an}(i)$ and $P_{ah}(i)$ ($i = 0, 1, ...N$) denote the acceptance probabilities for the new call and the handoff call requests, respectively, when the system is in state i.

To summarize, the QoS metrics P_B, P_D, and P_L are all functions of the CAC policy ϕ, defined by N, $P_{an}(i)$ and $P_{ah}(i)$, $i = 0, 1, ...N$.

The new call blocking probability is given by

$$P_B = \sum_{i=0}^{N} \pi(i)(1 - P_{an}(i)), \qquad (5.12)$$

the handoff dropping probability is

$$P_D = \sum_{i=0}^{N} \pi(i)(1 - P_{ah}(i)), \qquad (5.13)$$

and the packet loss probability caused by traffic congestion is

$$P_L = E(R_L(N)) = \sum_{i=0}^{N} \pi(i)R_L(i). \qquad (5.14)$$

5.2.2 Problem Formulation

The call admission control problem for VBR traffic is to decide the unknown variables N, $P_{an}(i)$ and $P_{ah}(i)$, which uniquely define a CAC policy such that the QoS metrics P_L, P_D and P_B are satisfied. We formulate the CAC problem as: to obtain the best media application quality by minimizing the P_L under certain constraints on the connection-level QoS. Mathematically, we would like to find the optimal CAC policy ϕ' such that

$$P_L(\phi') = minP_L(\phi),$$

subject to

$$P_B \leq P_B^{tar}, \text{ and } P_D \leq P_D^{tar}.$$

To simplify the problem, we consider a subset of the CAC schemes $\{\phi_s(N, T)\} \subset \{\phi\}$, where T is a real number, N is an integer, and $T \leq N$. The acceptance probabilities of new call requests (P_{an}) and handoff requests (P_{ah}) are

$$P_{an}(i) = \begin{cases} 1, & 0 \leq i < \lfloor T \rfloor \\ 0, & \lfloor T \rfloor + 1 \leq i \leq N \\ T - \lfloor T \rfloor, & i = \lfloor T \rfloor \end{cases} \qquad (5.15)$$

and

$$P_{ah}(i) = \begin{cases} 1, & 0 \le i < N \\ 0, & i = N \end{cases} \qquad (5.16)$$

In other words, the CAC policy admits a new connection request if and only if the current load is less than T, while admits a handoff request if and only if the current load is less than N. When T is a non-integer, the CAC policy admits a new connection request with the probability of the fractional part of T. This subset of CAC is called the limited fractional guard channel scheme, which is a superset of the guard-channel CAC with an integer value of T, and has been proved to be the optimal CAC policy under some constraints via the randomization of fractional T [1]. Within this subset, the problem of CAC schemes is to find the optimal values of N and T such that the QoS objectives are satisfied. In the following section, we propose a dynamic CAC scheme that adjusts the two parameters N and T adaptively to minimize the P_L while satisfying the given P_B and P_D constraints.

5.2.3   Minimize P_L Subject to P_B and P_D Constraints

Under a certain CAC policy $\phi_s(N, T)$ described in the above section, we can denote the P_B, P_D and P_L as functions of N and T, i.e., $P_B(N, T)$, $P_D(N, T)$ and $P_L(N, T)$, respectively. Apparently, $P_L(N, T) = 0$ when $N \le C$. Therefore, if there exists a T ($0 \le T \le C$) such that $P_B(C, T) \le P_B^{tar}$ and $P_D(C, T) \le P_D^{tar}$, then P_L could be minimized as zero. The CAC problem can thus be reduced to finding T that optimizes the connection-level QoS (including P_B and P_D). We consider the non-trivial case where $N \ge C$ such that $P_L(N, T) \ge 0$.

Based on the properties of $P_B(N, T)$, $P_D(N, T)$ and $P_L(N, T)$, the optimal stationary solution has been found [21]. An iteration algorithm has been proposed to obtain the optimal stationary CAC scheme for VBR traffic. However, different multimedia service types, such as video, audio, and data, may have different peak rates. Moreover, the traffic condition in terms of new/handoff call arrival and departure rates may be changing from time to time. Considering these practical aspects, the stationary optimal CAC schemes may not be efficient. Thus, dynamic schemes that take these factors into account are highly desirable.

For a cell with fixed channel capacity C, the P_L increases with the increase of admitted total traffic flows N. Based on this conjecture, in order to minimize the P_L, the dynamic CAC scheme should achieve minimum N and adjust T adaptively to keep P_B and P_D below their target constraints. Thus, the problem is equivalent to dynamic channel assignment as discussed in Section 5.1.3. It can be solved using the procedure described in Figure 5.4.

The dynamic CAC scheme is similar to the DCA algorithm proposed in Fig. 5.3. In line 27, the added condition ensures that N never falls below the channel capacity C. The reason is that when $N = C$, the $P_L = 0$. Further reducing N does not result in lower P_L, while the P_B and P_D increase. The initial N and T can be set to C. However, the performance of the dynamic CAC scheme converges faster with closer estimations to the optimal stationary values.

Note that in real-world wireless multimedia networks, the VBR traffic is generally more complex than the assumed independent identically distributed on-off model. In that case, although the formulation of P_L might be different from that in Equation (5.14), it is intuitive to infer that the same properties as those described above should apply. Thus our proposed scheme is applicable in real-world wireless multimedia networks with VBR traffics.

5.3 Simulation Results and Discussion

To evaluate the performance of the proposed dynamic call admission control and channel allocation schemes, we conducted simulations using the OPNET network simulator. The simulation results for CBR and VBR traffic are discussed below.

5.3.1 Simulation Results for CBR Traffic

Without loss of generality, we use a simple network model of a single cell with channel capacity $C = 100$. A number of call generators generate new call and handoff call requests from three different service classes: voice, audio and video, requiring r_1, r_2, and r_3 channels,

```
01  INITIALIZE N and T
02  SET TIMER
03  WAIT FOR CALL REQUEST ARRIVAL
04  If NEW CALL REQUEST ARRIVES
05     If (B_occupied + B_req) ≤ (N − T)
06        ADMIT NEW CALL WITH RATE B_req
07     Else
08        REJECT NEW CALL REQUEST
09        UPDATE P_B
10        If P_B ≥ P_B^tar
11           T = T + 1
12           If T > N
13              N = N + 1
14           RESET TIMER AND GO BACK TO 03
15  If HANDOFF CALL REQUEST ARRIVES
16     If (B_occupied + B_req) ≤ N
17        ADMIT HANDOFF CALL WITH RATE B_req
18     Else
19        REJECT HANDOFF REQUEST
20        UPDATE P_D
21        If P_D ≥ P_D^tar
22           If P_B ≥ P_B^tar OR T = 0
23              N = N + 1
24           Else
25              T = T − 1
26           RESET TIMER AND GO BACK TO 03
27  If TIMER EXPIRES AND N > C
28     N = N − 1
29     If T > N
30        T = N
31     RESET TIMER AND GO BACK TO 03
32  Else GO BACK TO 03
```

Figure 5.4: The proposed dynamic CAC scheme for minimizing P_L of VBR multimedia traffic with both P_B and P_D constraints.

Table 5.1: Traffic load parameter settings in OPNET simulation of dynamic CAC scheme for CBR traffic.

Application type (i)	r_i (channels)	λ_{ni} (call/sec)	λ_{hi} (call/sec)	η_i (1/sec)
voice (1)	1	0.6	0.1	0.02
audio (2)	2	0.3	0.05	0.02
video (3)	4	0.2	0.05	0.02

respectively, all with Poisson arrivals. The mean arrival rates of new call (handoff) requests from three classes are $\lambda_{n1}(\lambda_{h1})$, $\lambda_{n2}(\lambda_{h2})$, and $\lambda_{n3}(\lambda_{h3})$, respectively. The cell residence time is exponentially distributed with mean $\frac{1}{\eta_1}$, $\frac{1}{\eta_2}$, and $\frac{1}{\eta_3}$, respectively. The offered traffic load is defined as

$$L = \frac{1}{C} \sum_{i=1}^{3} \frac{\lambda_{ni} + \lambda_{hi}}{\eta_i} * r_i \tag{5.17}$$

Dynamic CAC: Minimize P_B Subject to P_D Constraint

In this section, the performance of the dynamic CAC scheme given in Fig. 5.1 is investigated.

First, we evaluate the effectiveness of the proposed dynamic call admission control scheme in achieving the objective by setting varying P_D^{tar} under a typical offered traffic load $L = 1.2$ with the parameters as in Table 5.1. Note all experimental results shown in this chapter are obtained by simulating a 20-hour network operation. Fig. 5.5 illustrates an example of dynamically changing threshold in our scheme where the P_D^{tar} was 0.01.

The performances in terms of three metrics, i.e., the P_D, the P_B and channel utilization U_C, are shown in Fig. 5.6. The nearly straight line in Fig. 5.6 (a) indicates that the proposed measurement-based dynamic guard channel scheme can always achieve the P_D^{tar} under this scenario. From Figs. 5.6 (b) and (c), we see that by increasing the P_D^{tar}, the P_B decreases, while the U_C increases accordingly.

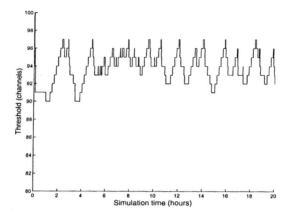

Figure 5.5: Illustration of the dynamic changing threshold.

Furthermore, we compared the performances of the proposed dynamic call admission control scheme with those of the traditional guard channel scheme under different traffic loads. Here, we consider two guard channel schemes with threshold T as 80 and 90 channels, which means 10 and 20 percent of channel capacity were exclusively reserved for handoff calls, respectively. The P_D^{tar} was set to 0.01. The actual P_D, the P_B, and U_C are illustrated in Fig. 5.7.

As shown in Fig. 5.7(a), the proposed dynamic guard channel scheme achieved the P_D^{tar} under varying traffic loads. However, both guard channel schemes cannot maintain the desired connection-level QoS in terms of the P_D^{tar}. Although they achieved a lower P_D under light traffic loads, they blocked more new calls than our dynamic scheme (see Fig. 5.7 (b)), thus resulting in lower U_C (see Fig. 5.7 (c)).

When traffic load increases to a certain extent, the handoff dropping probabilities for the fixed guard channel schemes exceeded the target and increased quickly. Although P_B is lower and U_C is higher than those of our scheme, they could not provide the primary connection-level QoS in terms of the P_D under heavy traffic loads.

Dynamic CAC: Minimize P_D Subject to P_B Constraint

In this section, the effectiveness of the proposed dynamic CAC scheme given in Fig. 5.2 in achieving the objective of different P_B^{tar} for multimedia traffic is investigated.

The network model has the same offered traffic load as in Table 5.1. The performances in terms of three metrics, *i.e.* the P_B, the P_D and the U_C, are shown in Fig. 5.8(a)-(c). Fig. 5.8(d) gives the time average of the dynamically changing threshold T.

Fig. 5.8 (a) shows that the proposed measurement-based dynamic guard channel scheme can achieve the P_B^{tar} when $P_B^{tar} \geq 0.2$ under this scenario. When $P_B^{tar} \leq 0.15$, the P_B cannot meet the constraint even if $T = C$ as shown in Fig. 5.8 (d). It can be seen that the lowest achievable P_B^{tar} is 0.172. We see from Figs. 5.8 (b) and (c) that the P_D and U_C decrease when P_B^{tar} increases. This is due to the decrease in the average of threshold T as shown in Fig. 5.8 (d).

DCA Subject to Joint P_B and P_D Constraints

In this section, we present simulation results of the DCA scheme proposed in Section 5.1.3. The same network model and traffic loads as those in the previous sections were used. The constraints were set as $P_B^{tar} = 0.1$ and $P_D^{tar} = 0.01$. Fig. 5.9 shows the simulation results of P_B and P_D. The corresponding dynamic values of C and T are illustrated in Fig. 5.10.

As shown in the figures, in the beginning with $C = 100$, the $P_B^{tar} \leq 0.1$ is not achievable. Thus, the DCA scheme increases C gradually to 133 until $P_B < P_B^{tar}$ when $T = C$. Then, T is decreased to lower P_D while maintaining P_B below the target. After both P_B and P_D are lower than their targets, C decreases accordingly to minimize the required channel resource. When the system reaches its stable status, both the P_B and the P_D converge to their target values. The average U_C achieved in this experiment is 84.5%.

Table 5.2 records simulation results of different P_B^{tar} and P_D^{tar} settings. The listed channel capacity C and threshold T are the average values over 20-hour simulation time. Compared to the stationary method proposed in [1], our dynamic scheme achieves a higher

Table 5.2: Simulation results of the DCA scheme.

Constraints		DCA Results				
P_B^{tar}	P_D^{tar}	C	T	P_B	P_D	U
0.05	0.01	138.812	135.553	0.04898	0.00999	0.815
0.05	0.05	131.598	131.452	0.04867	0.04868	0.848
0.1	0.01	126.342	121.068	0.09942	0.01165	0.845
0.1	0.05	120.521	119.051	0.09935	0.04993	0.871
0.2	0.01	111.340	103.381	0.19874	0.01009	0.853
0.2	0.05	104.710	101.445	0.19905	0.05034	0.890

resource utilization while both P_B and P_D are maintained close to the target. This is generally impossible for a stationary scheme. It demonstrates that the proposed DCA scheme is effective in minimizing the required channel capacity while providing connection-level QoS guarantees.

5.3.2   Simulation Results for VBR Traffic

In the following, we examine the performance of the proposed dynamic CAC algorithm for VBR traffic for multiple service types.

A network model for VBR traffic was built in OPNET simulator to simulate a single cell with channel capacity $C = 20$ in wireless multimedia networks. Each channel has a data rate of 9600 bits/sec. The VBR traffic model was built as an on-off model with configurable packet generation parameters listed in Table 5.3

The network traffic load configurations are listed in Table 5.4. The r_i for each service type is the weighting factor of the peak rate during the on-period. It is used by the above VBR on-off model to scale the packet inter-arrival time generated. Thus, the offered traffic load as defined in 5.3.1 is

$$L = \frac{1}{C} \sum_{i=1}^{3} \frac{\lambda_{ni} + \lambda_{hi}}{\eta_i} * r_i = 2.5$$

The constraints of P_B and P_D are given as $P_B^{tar} = 0.1$ and $P_D^{tar} = 0.01$. For comparison purpose, we found the optimal corresponding

Table 5.3: Configuration of the multimedia VBR traffic model parameters in the experiment of dynamic CAC scheme.

Parameter	Value
On-state time distribution	Exponential with mean 1 sec
Off-state time distribution	Exponential with mean 3 sec
Packet inter-arrival time distribution during on-state	Exponential with mean 0.1 sec
Packet size distribution	Constant with mean 120 bytes

Table 5.4: Traffic load parameter settings in OPNET simulation of dynamic CAC scheme for VBR traffic.

Application type (i)	r_i (channels)	λ_{ni} (call/sec)	λ_{hi} (call/sec)	η_i (1/sec)
voice (1)	1	0.2	0.05	0.01
audio (2)	2	0.05	0.015	0.01
video (3)	4	0.025	0.005	0.01

Table 5.5: Comparison of simulation results between stationary and dynamic CAC schemes for VBR traffic.

Schemes	P_B	P_D	P_L	N	T
Stationary CAC	0.1002	0.009	0.00245	56	53.1
Dynamic CAC	0.0943	0.010	0.00213	55.3	52.2

stationary CAC schemes with $N = 56$ and $T = 53.1$ by trial-and-error experiments, so that the P_B and P_D are closest to the targets. Then we compared the P_L with that of the dynamic CAC scheme. Simulation results for the two schemes are summarized in Table 5.5, where N and T have been averaged over the simulation time. To illustrate the dynamic CAC scheme, the variation of N and T values are shown in Fig. 5.11.

It can be shown that the proposed dynamic CAC scheme for VBR multimedia traffic can achieve a lower P_L than the stationary ones, while satisfying both the P_B and the P_D constraints.

5.4   Conclusion and Future Work

In this chapter, we described dynamic CAC and resource allocation schemes to provide both connection-level and packet-level QoS in wireless multimedia networks.

For CBR multimedia traffic whose packet-level QoS is based on connection-level QoS, different objectives of connection-level QoS provisioning were formulated. First, a dynamic CAC scheme was proposed to provide the connection-level QoS guarantee by achieving the minimum connection blocking probability subject to the constraint on the handoff dropping probability. This scheme adopts a novel strategy called prompt-decreasing/timer-increasing (PDTI) to dynamically adjust the threshold for handoff channel reservation. It can maintain the handoff dropping probability at a target rate predefined in the system specification, while maximizing resource utilization and minimizing the new call blocking probability. Second, the proposed dynamic guard channel CAC scheme was extended to provide the minimum handoff dropping probability under the constraint on the new call blocking probability. Third, a dynamic channel assignment scheme was developed to find the minimum channel capacity of a cell with both the new call blocking and the handoff dropping probability constraints. By adjusting the necessary capacity and the threshold dynamically, it provides a guide for capacity allocation to a cell in a wireless multimedia network with connection-level QoS provisioning. This scheme could also be used to provide joint optimization of connection-level and packet-level QoS (in terms of packet loss probability).

Simulations were carried out using the OPNET network simulator to prove the efficiency and effectiveness of our proposed schemes to achieve various objectives. As measurement-based dynamic schemes, the proposed schemes are not restricted by any traffic pattern and are applicable to varying conditions found in wireless networks. The low complexity features enable their practical deployment in real-world applications.

As stated in Section 5.2, the dynamic CAC scheme proposed for VBR traffic was developed using the congestion-caused packet loss probability as packet-level QoS metric. It is however also applicable to other packet-level QoS metrics with similar properties, for example, delay and packet loss at other layers. It is reasonable to assume that the performance of these packet-level QoS metrics decreases when the number of connections in the system increases. Thus, it should be interesting to extend the developed schemes for other aspects of packet-level QoS.

To summarize, the work in this chapter provides some foundation for the joint optimization of packet-level and connection-level QoS in the future packet-switching wireless multimedia networks. Further research with more complex yet practical considerations based on our current work should be of great commercial as well as academic value.

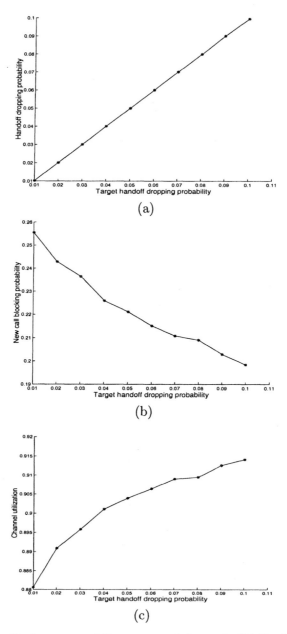

Figure 5.6: Performance of the proposed dynamic CAC scheme with handoff dropping probability constraint: (a) the handoff dropping probability, (b) the new call blocking probability, and (c) channel utilization.

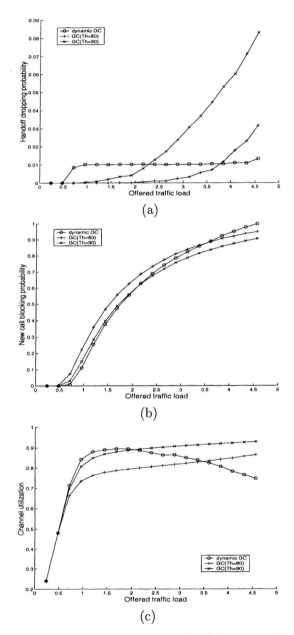

Figure 5.7: Performance comparison under different traffic load: (a) channel utilization, (b) the new call blocking probability, (c) the handoff dropping probability.

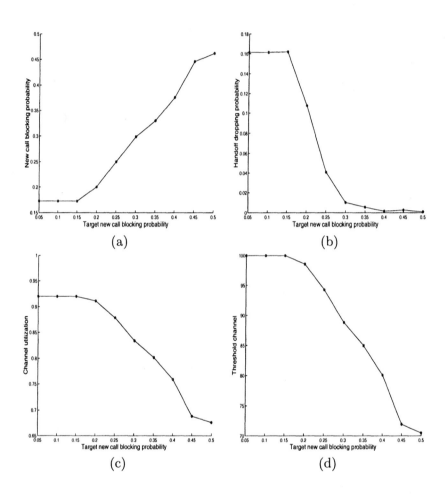

Figure 5.8: The performance of the proposed dynamic CAC scheme with the new call blocking probability constraint: (a) the new call blocking probability, (b) the handoff dropping probability,(c) channel utilization, and (d) time average of threshold T

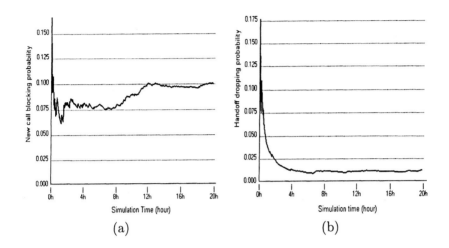

Figure 5.9: The performance of the proposed DCA scheme: (a) the new call blocking probability and (b) the handoff dropping probability.

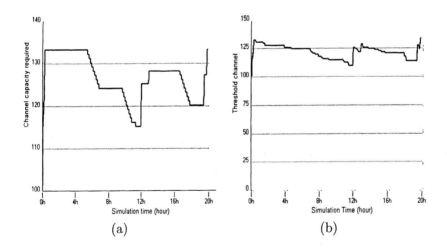

Figure 5.10: The dynamic behavior of (a) the channel capacity C and (b) the threshold T for the proposed DCA scheme.

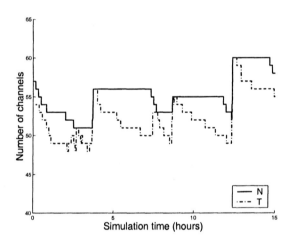

Figure 5.11: Illustration of varying N and T for the dynamic CAC scheme for VBR multimedia traffic.

BIBLIOGRAPHY

[1] R. Ramjee, D. Towsley, and R. Nagarajan, "On optimal call admission control in cellular networks," *Wireless Networks*, 3(1):29–41, May 1997.

[2] T. Kwon, S. Kim, Y. Choi, and M. Naghshineh, "Threshold-type call admission control in wireless/mobile multimedia networks using prioritised adaptive framework," *Electronics Letters*, 36(9):852–854, Apr 2000.

[3] J. M. Peha and A. Sutivong, "Admission control algorithms for cellular systems," *Wireless Networks*, 7(2):117–125, Apr 2001.

[4] W. S. Soh and H. S. Kim, "Dynamic guard bandwidth scheme for wireless broadband networks," In *Proc. of the IEEE INFOCOM 2001*, volume 1, pages 572–581, Apr 2001.

[5] M. Naghshineh and S. Schwartz, "Distributed call admission control in mobile/wireless networks," *IEEE Journal on Selected Areas in Communication*, 14(4):711–717, May 1996.

[6] S. Wu, K. Y. M. Wong, and B. Li, "A new distributed and dynamic call admission policy for mobile wireless networks with qos guarantee," In *Proc. of the Ninth IEEE International Symposium on Personal, Indoor and Mobile Radio Communications*, volume 1, pages 260–264, Sept. 1998.

[7] T. Kwon and Y. Choi, "Call admission control for mobile cellular network based on macroscopic modeling of vehicular traffic," In *Proc. of IEEE VTC*, pages 1940–1944, May 1998.

[8] D. Hong and S. S. Rapport, "Traffic modeling and performance analysis for cellular mobile radio telephone systems with prioritized and non-prioritized handoff procedures," *IEEE Trans. on Vehicular Technology*, 35(3):77–92, Aug. 1986.

[9] S. H. Oh and D. W. Tcha, "Prioritised channel assignment in a cellular radio network," *IEEE Trans. on Communications*, 40(7):1259–1269, Jul. 1992.

[10] S. Jordan, "Resource allocation in wireless networks," *Journal of High Speed Networks*, 5(1):23–34, 1996.

[11] J. Tajima and K.Imamura, "A strategy for flexible channel assignment in mobile communication systems," *IEEE Trans. on Vehicular Technology*, 37(2):92–103, 1988.

[12] M. Zhang and T.-S. Yum, "The nonumiform compact pattern allocation algorithm for cellular mobile systems," *IEEE Trans. on Vehicular Technology*, 40(2):387–391, 1991.

[13] J. Zander, "Generalized reuse partitioning in cellular mobile radio," In *Proc. IEEE Veh. Technol. Conf.*, pages 181–184, 1993.

[14] A. K. Talukdar, B. R. Badrinath, and A. Acharya, "Integrated services packet networks with mobile hosts: Architecture and performance," *Wireless Networks*, 5(2):111–124, Mar. 1999.

[15] A. K. Talukdar, B. R. Badrinath, and A. Acharya, "MRSVP: a resource reservation protocol for an integrated services network with mobile hosts," *Wireless Networks*, 7(1):5–19, Jan 2001.

[16] H. Liu and M. E. Zarki, "Transmission of video telephony images over wireless channels," *Wireless Networks*, 2(3), Aug 1996.

[17] S. Kaiser, "OFDM code-division multiplexing in fading channels," *IEEE Trans. on Communications*, 50(8):1266 –1273, Aug 2002.

[18] S. Siwamogsatham and M. P. Fitz, "Robust space-time codes for correlated Rayleigh fading channels," *IEEE Transactions on Signal Processing*, 50(10):2408 –2416, Oct 2002.

[19] V. Bharghavan, A. Demers, S. Shenker, and L. Zhang, "MACAW: a media access protocol for wireless LANs," In

ACM SIGCOMM Computer Communication Review, Proc. of the conf. on Communications architectures, protocols and applications, volume 24, pages 212–225, London, United Kingdom, Oct 1994.

[20] J. J. Garcia-Luna-Aceves and A. Tzamaloukas, "Receiver-initiated collision avoidance in wireless networks," *Wireless Networks*, 8(2/3):249–263, Mar 2002.

[21] L. Huang, "Call Admission Control and Resource Allocation for QoS Support in Wireless Multimedia Networks," *Ph.D Thesis, University of Southern California*, May 2003.

Chapter 6

HANDOFF SCHEMES IN TDMA/FDMA SYSTEMS

This chapter addresses the issue of how to provide seamless handoff to mobile users, under the constraint of limited resources, in a multimedia wireless network. We adopt the concept of the guard channel (GC) scheme, which gives preferential treatment to the high priority calls, including handoff calls, by reserving a fixed number of channels exclusively for them. However, such schemes decrease the handoff dropping rate at the cost of increasing the blocking rate for other users due to poor channel utilization. To deal with this challenge, we introduce a dynamic resource reservation module to efficiently estimate the resources needed to be reserved for high priority calls, by using the SNR and distance information of a mobile in the neighboring cells.

The rest of this chapter is organized as follows. In Section 6.1, we present an extension of the conventional fixed GC scheme with multiple thresholds to accommodate multiple priority classes traffic in multimedia wireless networks. By a fixed GC scheme, we mean that the number of guard channels in a cell is fixed during the entire operation process. In order to achieve a higher degree of system utilization, we develop dynamic GC schemes in later sections. In Section 6.2.1, we describe a framework that characterizes interaction among communication elements, and find criteria for good CAC and RR designs to improve the QoS measure. A service model and the corresponding application profiles are detailed in Section 6.2.2, followed by CAC and RR designs in Section 6.2.3 along with the discussion of the rationale behind them. Sections 6.3.2 and 6.3.3 provide some simulation results. Finally, concluding remarks and future research work are given in Section 6.3.4.

Please note that the schemes presented in Chapters 3 to Chapter 5 also addressed the issues of CAC and handoff. However, the schemes presented in this chapter have somewhat different approach as they use different service model and traffic profiles.

6.1 Fixed Guard Channel Handoff Scheme with Multiple Thresholds

A wireless multimedia system cannot always meet different QoS requirements of mobile users due to resource constrains. Therefore, the system requires rules to decide who will receive the services according to predefined cost function(s), to avoid unwanted call blocking and handoff dropping while maximizing channel utilization. Resource reservation and call admission schemes should be integrated with the handoff mechanism to provide more flexibility to all mobile users and better QoS guarantees for premium users. Many different admission control strategies have been discussed in the literature to provide priorities to high-priority call and handoff requests without undue blocking of new connection requests. We extend the simple GC scheme to multiple thresholds in this section.

6.1.1 Service Model

The service model and the associated application profiles describe the characteristics of the traffic. Three attributes are included in the application profile of call i: (1) the bandwidth Consumption Φ_i, (2) the handoff or the new call indicator denoted by $I_{h,n}^i$, and (3) the priority class Π_i. In a mobile communication system with a maximum capacity of N channels, the ith $(i < N)$ user's application profile can be represented by $\Im\{\Phi_i, I_{h,n}^i, \Pi_i\}$. The communication entity and the resource reservation control mechanism take necessary action according to the information in the application profile.

The bandwidth consumption Φ_i is different for different media types in typical applications. Currently, we include both audio and video in our media and their values are assigned accordingly. Parameter $I_{h,n}^i$ is a binary number to indicate whether the call is a handoff or a new call request. The priority class attribute, denoted by Π_i, allows users to request one of the two priority service classes, i.e. the high and the low priorities. Users can request a connection of one of the two multimedia traffic types with one of the two different priority attributes (high or low). The indicator $I_{h,n}^i$ is automatically assigned by the wireless system according to each call connection status. While making a decision on admitting a low priority call, the network has to reserve the resources for the expected incoming traffic of a higher priority within the period of its residential time. The residential time means the period between the time the call arrives at a cell (or a new call born in this cell) to the time when the call completes connection in this cell (or handoff to other cell).

6.1.2 Proposed Multiple Thresholds GC Scheme

Let us first assign the priority order as follows: handoff call with high priority class (highest priority), new call with high priority class, handoff call with low priority class, and new call with low priority class (lowest priority).

The concept of the proposed multiple threshold GC scheme is illustrated in Fig. 6.1. In Fig. 6.1, the number of reserved channels, i.e. guard channels, is determined by the thresholds set for different priority classes. For each call request, the call admission control (CAC) mechanism calculate available resource (Not allocated resource in Fig. 6.1) according to their priority level.

The call admission control decision rule is illustrated in Fig. 6.2. The N_{busy} indicate the number of current busy channels. The associated GC thresholds are set to TH1=0, TH2=m_1, TH3=m_2 and TH4=m_3 for high to low priority classes, respectively.

6.1.3 Analytical Model

We can model multiple GC schemes by a M/M/C/C queueing system as in Fig. 6.3. In this model, we consider a traffic scenario

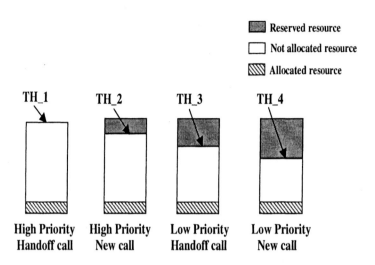

Figure 6.1: Illustration of multiple thresholds applied to different priority classes.

with differential treatments among priority classes. Four priority classes, high priority handoff call, high priority new call, low priority handoff call, and low priority new call are considered. The bandwidth consumption for each call is considered equally. In order to provide preferential treatment to higher priority call, the associated GC thresholds are set to TH1=0, TH2=m_1, TH3=m_2 and TH4=m_3, respectively. The value of m_1 to m_3 are in number of channels.

We assume the Poisson distribution for total call arrivals (including all priority classes) with an average rate of λ and exponentially distributed call holding time T_c with a mean $E[T_c] = 1/\mu$. We also assume that the cell residence time is T_r for a mobile user which is exponentially distributed with a mean $E[T_r] = 1/\eta$. The average channel occupation time is thus $1/(\mu + \eta)$ [2]. The call arrival rates, average call holding time, and average cell residence time for each priority class are denoted as $\alpha_j \cdot \lambda$, $1/(\alpha \cdot \mu)$ and $1/(\alpha \cdot \eta)$, respectively. Here, $j \in \{1, 2, 3, 4\}$ for priority classes from high to low,

Call Arrival

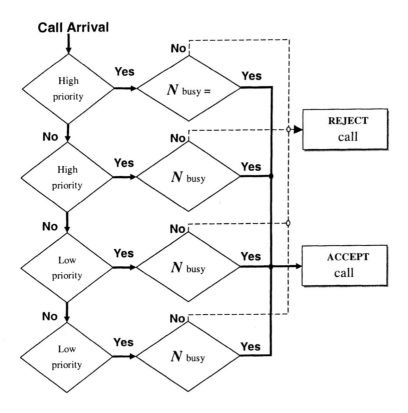

Figure 6.2: The decision rule for call admission control.

respectively, and $(\alpha_1 + \alpha_2 + \alpha_3 + \alpha_4) = 1$. The call blocking rate for the priority classes are given by Eqs. (6.4)-(6.7).

Then, we can derive the steady-state probability P_j for j channels to be busy as follows.

$$\lambda P_0 = (\mu + \eta) P_1$$
$$\lambda P_0 + 2 \cdot (\mu + \eta) P_2 = \lambda P_1 + (\mu + \eta) P_1$$

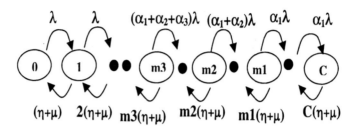

Figure 6.3: The queueing model for the multiple-threshold GC scheme.

$$\lambda P_{m_3-1} + (m_3+1)(\mu+\eta)P_{m_3+1} = (\alpha_1+\alpha_2+\alpha_3)\lambda P_{m_3} + (m_3)(\mu+\eta)$$

$$\vdots$$

$$(\alpha_1+\alpha_2+\alpha_3)\lambda P_{m_2-1} + (m_2+1)(\mu+\eta)P_{m_2+1} =$$
$$(\alpha_1+\alpha_2)\lambda P_{m_2} + (m_2)(\mu+\eta)P_{m_2}$$

$$\vdots$$

$$(\alpha_1+\alpha_2)\lambda P_{m_1-1} + (m_1+1)(\mu+\eta)P_{m_1+1} = \alpha_1\lambda P_{m_1} + (m_1)(\mu+\eta)P_m$$

$$\vdots$$

$$\alpha_1\lambda P_{m_1-1} + (m_1+1)(\mu+\eta)P_{m_1+1} = \alpha_1\lambda P_{m_1} + (m_1)(\mu+\eta)P_{m_2}$$

$$\vdots$$

$$\alpha_1\lambda P_{C-1} = C(\mu+\eta)P_1$$

$$(\bullet$$

The above equation set can be summarized below.

$$P_j = \begin{cases} \frac{1}{j!}\cdot(\frac{\lambda}{\mu+\eta})^j\cdot P_0 & \text{if } 0\leq j\leq m_3 \\ \frac{(\alpha_1+\alpha_2+\alpha_3)^{j-m_3}}{j!}\cdot(\frac{\lambda}{\mu+\eta})^j\cdot P_0 & \text{if } m_3\leq j\leq m \\ \frac{(\alpha_1+\alpha_2)^{j-m_2}\cdot(\alpha_1+\alpha_2+\alpha_3)^{m_2-m_3}}{j!}\cdot(\frac{\lambda}{\mu+\eta})^j\cdot P_0 & \text{if } m_2\leq j\leq m \\ \frac{(\alpha_1)^{j-m_1}\cdot(\alpha_1+\alpha_2)^{m_1-m_2}\cdot(\alpha_1+\alpha_2+\alpha_3)^{m_2-m_3}}{j!}\cdot(\frac{\lambda}{\mu+\eta})^j\cdot P_0 & \text{if } m_1\leq j\leq C \end{cases}$$

$$(6.2)$$

where

$$P_0 = \left(\begin{array}{l} \sum_{j=0}^{m_3} \frac{1}{j!} \cdot (\frac{\lambda}{\mu+\eta})^j + \sum_{j=m_3+1}^{m_2} \frac{(\alpha_1+\alpha_2+\alpha_3)^{j-m_3}}{j!} \cdot (\frac{\lambda}{\mu+\eta})^j + \\ \sum_{j=m_2+1}^{m_1} \frac{(\alpha_1+\alpha_2)^{j-m_2} \cdot (\alpha_1+\alpha_2+\alpha_3)^{m_2-m_3}}{j!} \cdot (\frac{\lambda}{\mu+\eta})^j + \\ \sum_{j=m_1+1}^{C} \frac{(\alpha_1)^{j-m_1} \cdot (\alpha_1+\alpha_2)^{m_1-m_2} \cdot (\alpha_1+\alpha_2+\alpha_3)^{m_2-m_3}}{j!} \cdot (\frac{\lambda}{\mu+\eta})^j \end{array} \right)^{-1}$$

$$(6.3)$$

Based on (6.2) and (6.3), we can derive the call blocking rate for each priority class as given by equations from (6.4) to (6.7). The blocking rate for the high priority handoff call, the high priority new call, the low priority handoff call and the low priority new call are denoted by P_{c_1}, P_{c_2}, P_{c_3} and P_{c_4}, respectively.

$$P_{c_1} = P_C \tag{6.4}$$

$$P_{c_2} = \sum_{j=m_3}^{C} P_j \tag{6.5}$$

$$P_{c_3} = \sum_{j=m_2}^{C} P_j \tag{6.6}$$

$$P_{c_4} = \sum_{j=m_1}^{C} P_j \tag{6.7}$$

6.2 Dynamic Guard Channel Handoff Scheme

The dynamic GC scheme achieves a better performance by exploiting the information of message loops. Message loops are internetworking communications among wireless system components, i.e. Mobile terminals (MT), Base stations (BS), and Main telephone switching office (MTSO). In the following sections, we will introduce mechanisms of these message loops in a mobile simulation system, and then propose a dynamic scheme.

6.2.1 Mobile Simulation System

Message Loops in the Mobile Simulation System

A mobile network consists of three key communication elements: the mobile terminal (MT), the base station (BS) and the main telephone switching office (MTSO) . Their interactions are described in Fig. 6.4 to Fig. 6.7 by message loops.

- **NCQ-ML:**

  The *New Call Request Message Loop* (see Fig. 6.4) is responsible for requesting a new call. It consists of four messages: *Request_New*, *Forward_Req_New*, *Resp_Req* and *Forward_Resp_Req*. Actions will be taken by each communication element depending on the message they receive. Each mobile terminal sends a new call request to its BS, according to its application profile at the startup. Finally, mobile terminal (MT) will either receive an accept or reject acknowledge. MT starts sending data when it receives an accept acknowledge and terminates the connection request if it receives a reject acknowledge. BS forwards a new call request to MTSO. Call admission control module in the MTSO will then decide whether to admit the request according to the application profile of that call request, its current cell capacity as well as predefined cost function.

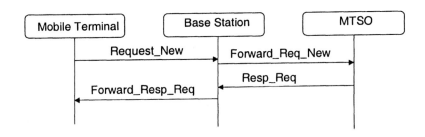

Figure 6.4: The new call request message loop in the simulation system.

- **UHC-ML:**

  The *Update Handoff Candidate Message Loop* (see Fig. 6.5) updates the handoff candidates and consists of *Call_Update* and *Forward_Call_Update* messages. This message loop begins with an active MT, whose signal is detected by neighboring BSs. Each BS receives signals not only from its associated

Table 6.1: Message in New Call Request Message Loop

Message	From → To	Description
Request_New	MT → BS	New call request send to BS
Forward_Req_New	BS → MTSO	Forward the new call request
Resp_Req	MTSO → BS	MTSO's decision feedback
Forward_Resp_Req	BS → MT	MTSO's decision feedback via BS to MT

MTs but also from MTs in the neighboring cells within its power range. If received SNR of an MT in a neighboring cell is greater than the threshold, BS will send a report message to MTSO to register itself as the handoff candidate of that neighboring MT. Mobile terminals can know where to handoff by referencing this handoff candidate registration (HCR) table in the MTSO when their signal fades.

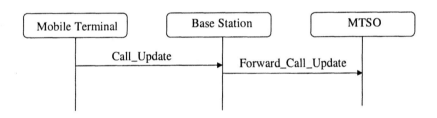

Figure 6.5: The update handoff candidate message loop in the simulation system.

- **HR-ML:**
  The *Handoff Request Message Loop* (see Fig. 6.6) consists of *Call_Update, Handoff_Request, Resp_Hoff_Request* and *Forward_Resp_Hoff_Req* messages to request a handoff. The BS makes a handoff request to MTSO when it finds that its associated MT has signal degraded down below the threshold.

Table 6.2: Message in Update Handoff Candidate Message Loop

Message	From → To	Description
Call_Update	MT → BS	MT send out signals to allow its current BS and neighboring BS within its radio scope to update handoff candidates
Forward_Call_Update	BS → MTSO	If the signal of *Call_Update* is stronger than "reporting" threshold, such BS will send out message to register itself as handoff candidate

MTSO then checks the HCR table to find out a good candidate cell and admits the handoff request. The decision is sent back to the new as well as old base station, which is forwarded to MT.

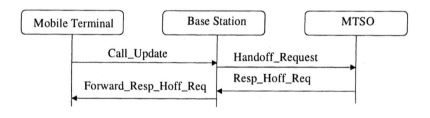

Figure 6.6: The handoff request message loop in the simulation system.

- **CT-ML:**
  The *Call Terminate Message Loop* (see Fig. 6.7) includes *Call_Term* and *Forward_Call_Term* messages to terminate calls.

Table 6.3: Message in Handoff Request Message Loop

Message	From → To	Description
Call_Update	MT → BS	MT send out signals to allow current BS and neighboring BS within its radio scope to update handoff candidates
Handoff_Request	BS→ MTSO	If the signal of *Call_Update* is weaker than acceptable threshold, such BS will send out message to request handoff
Resp_Hoff_Req	MTSO → BS	MTSO's feedback for handoff request
Forward_Resp_Hoff_Req	BS → MT	MTSO's feedback for handoff request

Use of SNR and Distance Information

As mentioned earlier, a handoff candidate registration (HCR) table is maintained in the MTSO to record the handoff candidate for each MT. This is how a mobile terminal knows where to handoff when its signal fades. We utilize the SNR and the associated distance information [3] of the mobile radio propagation to identify the weighting of resources to be set aside for high-priority calls. The relation between the SNR and the associated distance are given below.

$$P_r = P_t - 157.7 - 38.4 \log(d) + 20 \log(h_1) + 10 \log(h_2) \quad (6.8)$$
$$+ G_t + G_m,$$
$$SNR = P_n - P_r \quad (6.9)$$

where P_t and P_r are the transmitted and received power (in decibels), respectively, P_n is the environment noise (in dB) received by BS, and G_t and G_m are antenna gains in dB for transmitter and mobile,

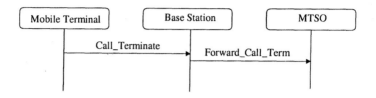

Figure 6.7: The call terminate message loop in the simulation system.

Table 6.4: Message in Call Terminate Message Loop

Message	From → To	Description
Call_Terminate	MT → BS	MT receive reject response from MTSO feedback message or upon call termination itself
Forward_Call_Term	BS → MTSO	Call termination message is forwarding to MTSO

respectively. The antenna heights of h_1 (BS) and h_2 (MT) are in feet. Suggested values are given in Table 4.1 in [3], for h1=100ft, h2=10ft. Finally, d is the distance, measured in miles, for a given received power. According to above formulae, the SNR and the distance values can be exchanged with each other.

6.2.2   Service Model and Application Profile

Service Model with Multiple QoS Classes

We consider a more comprehensive service model for dynamic scheme here than that used in Section 6.1.1, with the following service attributes:

- (Φ_{max}, Φ_{min}):
  The *maximum and minimum bandwidth requirements* describe the bandwidth consumption of the traffic. If traffic is non-rate

adaptive, the value of Φ_{max} is equivalent to Φ_{min}. On the other hand, if traffic is rate adaptive, the value of Φ_{min} should set to be less than that of Φ_{max}.

- I_{rate}:
  The *rate adaptivity indicator* describes whether a connection is flexible in its bandwidth requirements. If a connection is rate adaptive, it can be serviced in a degraded mode when congested, and it thus has high probability to receive service, either in full or degraded rate. The value of I_{rate} is give by

$$I_{rate} = \begin{cases} 1 & \text{if call is rate adaptive} \\ 0 & \text{if call is non-rate adaptive} \end{cases} \tag{6.10}$$

- Π:
  The *priority* with a high value is assigned to connections that are willing to pay more. They are likely to receive better QoS guarantees in terms of better chance to receive service and in better quality mode. Similarly, system will gain higher rewards if it services such priority calls. The reward function is defined in Eq. (ch6eq:rewardFunction).

There is another traffic attribute, i.e. mobility, denoted as M. It describes the speed property of a mobile terminal. Different mobility traffic will have different weighting factor of the estimated bandwidth needed to be reserved. This information is taken into consideration in the process of resource reservation estimation.

In a mobile communication system with a maximum capacity of N channels, the i-th, $(i < N)$ user's application profile, $\Im$, can be represented as

$$\Im_i = \{(\Phi_{max,i}, \Phi_{min,i}), \Pi_i, M_i\}. \tag{6.11}$$

While making a decision on admitting a low priority call, the network has to reserve resources for expected incoming higher-priority traffics in the neighboring cells within the range of a BS's awareness. The application profile of each active call carries its own connection description for the use of requesting a connection as well as the priority referencing by other calls.

6.2.3   Proposed Dynamic Guard Channel Scheme

Proposed Resource Reservation Estimation

As mentioned in Section 6.2.1, BS will register itself as a handoff candidate when it receives signals from a specific MT whose SNR is higher than a threshold. Traditionally, this handoff candidate registration (HCR) table is used to inform the MT about where to handoff when its signal fades. This table also provides very useful information to estimate future handoff calls for a given cell. Our proposed dynamic resource reservation estimator is based on a non-linear weighting sum as described below, which is different from Linear Weighted Sum (LWS) scheme described in previous Section.

- When an incoming call j requests for resources in the cell j_{target_cell}, a non-linear weighting curve as shown in Fig. 6.8 is used to estimate the threshold of resources to be reserved in this cell for serving higher-priority active calls from set S. The following two equations define the weighting factor with and without traffic mobility differentiation, respectively.

$$W_i = \begin{cases} d_{Th}/d_i & \text{if } d_i > d_{Th} \\ 1 & \text{if } d_i < d_{Th} \end{cases} \qquad (6.12)$$

  where W_i is the weighting factor and d_i is the distance obtained by (6.9) and (6.9), and

$$W_i = \begin{cases} T_{Th}/T_i & \text{if } T_i > T_{Th} \\ 1 & \text{if } T_i < T_{Th} \end{cases} \qquad (6.13)$$

  where W_i is the weighting factor when mobility is considered in the service model, and T_i and T_{Th} are time related parameters.

- Φ_{res} as defined in (6.18) represents the resources needed to be reserved for the use of priority calls in our proposed CAC algorithm. Let us first define a set $S(j)$ for call j to be considered

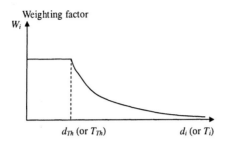

Figure 6.8: The non-linear weighting curve for resource reservation estimation.

for the estimation of resource reservation. The set $S(j)$ consists of all neighboring active calls that satisfy the following two criteria. First, the handoff candidate cell of call i in HCR table is the same as the target cell of call j. Second, the priority of call i is higher than that of incoming call j. Note that, there is one more hidden assumption, i.e. the current cell of neighboring call i is not equal to the target cell of incoming call j. We define the following operations for call i (similar for call j)

$$\Pi(i) \quad : \quad \text{Priority of call } i \qquad\qquad (6.14)$$
$$\Lambda(i) \quad : \quad \text{The handoff candidate cell of call } i \quad (6.15)$$
$$\Lambda^*(i) \quad : \quad \text{The target cell of call } i \qquad\qquad (6.16)$$
$$\text{(the cell with maximum SNR among}$$
$$\text{the handoff candidate cells)}$$

Thus, set $S(j)$ can be represented as

$$S(j) = \{i | \Pi(i) > \Pi(j), \Lambda(i) = \Lambda^*(j)\}. \qquad (6.17)$$

Fig. 6.9 illustrates an example of set S. In this example, we consider base station BS_0 as the target cell, on which call admission control algorithm and resource reservation estimation scheme are applied. Assume that at an instance, there is an incoming call, call j, who just broke the connection with the home base station, BS_2, and making a handoff call request to target cell of BS_0. Apparently, incoming call j knows that BS_0 is its target cell because call j has got the handoff direction command from MTSO. MTSO has scanned the handoff candidate registration table associated to the call j and find out that BS_0 can receive the strongest signal strength from call j. Upon the call request of j, who has low priority class attribute, resource reservation and call admission control mechanism will do appropriate resource reservation for the potential high priority calls in the neighboring cells. In our case, we examine the operations in Eq. (6.15 and (6.15) on current neighboring active call i_1 to i_5 one by one to decide which elements are considered in set S.

- **Operation I:** $\Pi(i_2), \Pi(i_3), \Pi(i_5) > \Pi(j)$.
  Among active calls, i.e. i_1 to i_5, in the neighboring cells, only i_2, i_3 and i_5 whose priority greater than incoming handoff call request j.

- **Operation II:** $\Lambda^*(j) = BS_0$; $\Lambda(i_1), \Lambda(i_3), \Lambda(i_4), \Lambda(i_5) = BS_0$.
  Check the handoff candidate registration (HCR) table for each neighboring call, we found calls i_1, i_3, i_4 and i_5 having BS_0 on their lists.

- **Set S:** $SetS = \{i_3, i_5\}$.
  According the definition, set S includes neighboring calls who satisfy both Operations I and II.

• Two types of mobility - high and low - are considered in our simulation. We assign 1 unit and 2 unit speeds for low and high mobility traffic, respectively. The time related parameters, $\{T_i, T_{Th}\}$, in Eq. (6.13) can be represented as $\{d_i, d_{Th}\}$ and $\{d_i/2, d_{Th}\}$ for low and high mobility traffic, respectively.

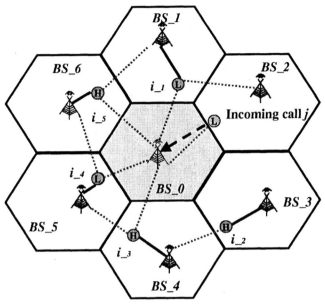

H Terminal with high priority class i_n ◀┄   Call request association

L Terminal with low priority class i_n ┄┄┄┄   Handoff candidate association

🗼 Base Station (BS)   ▬▬▬   Actual connection association

Figure 6.9: Illustration of set S

This implies that a high speed MT is more likely to handoff into current cell even when it is farther from a low speed MT.

$$\Phi_{res}(j) = \sum_{i \in S} \omega_i \cdot \Phi_{min}(i). \qquad (6.18)$$

- The HCR table helps BS in finding out the most likely potential handoff calls, which would help increase the resource reservation efficiency.

Proposed Call Admission Control (CAC) Algorithm

Our proposed CAC algorithm for new call request is illustrated as follows:

> **If** Call request is non-rate adaptive
>> **If** $(N_{busy} + \Phi_{max}) < (C - \Phi_{res})$
>>> Admit call request with bandwidthΦ_{max}
>> **Else** Reject call request
>> **End**
>
>> **Else** / * *It is a rate adaptive call* * /
>>> **If** $(N_{busy} + \Phi_{max}) < (C - \Phi_{res})$
>>>> Admit call request withΦ_{max}
>>> **Else If** $(N_{busy} + 1/2 \cdot \Phi_{max}) < (C - \Phi_{res})$
>>>> Admit call request with $1/2 \cdot \Phi_{max}$
>>> **Else If** $(N_{busy} + \Phi_{min}) < (C - \Phi_{res})$
>>>> Admit call request withΦ_{min}
>>> **Else** Reject call request
>>> **End**
> **End**

The notations are summarized as in Table 6.5.

Table 6.5: Parameters used in call admission control algorithm

Parameter	Legend
C	Total number of channels
N_{busy}	Number of busy channels
Φ_{res}	Estimated resource needed to be reserved
Φ_{max}	Maximum number of channels requested
Φ_{min}	Minimum number of channels requested

Finally, we define our reward function as

$$R = \sum_{accept\ call\ i} \Phi_{admit}(i) \cdot \omega_{h,n}(i) \cdot \omega_\Pi(i), \qquad (6.19)$$

where $\Phi_{admit}(i)$ is the bandwidth usage admitted in the process of CAC algorithm test for call i. $\omega_\Pi(i)$ is the reward weighting factor for priority call i. Similarly, $\omega_{h,n}(i)$ is the reward weighting factor for handoff or new call requests.

6.3 Simulation Results and Discussion

6.3.1 Simulation Results for Fixed Guard Channel Handoff Scheme with Multiple Thresholds

Two most popular forms of network simulation are the analytical modelling simulation (AMS) and the discrete event simulation (DES). AMS characterizes the network as a set of equations by exploring queuing theory. An over-simplified analytical model often leads to unrealistic simulation results. DES characterizes the network as a set of finite state machines (FSM) according to control techniques. The nature of random processes and dynamic evolution of FSM make it possible to simulate more complicated and realistic network control mechanisms. While DES has many advantages, it requires far greater processing time. Fortunately, with the advance in the computing power, DES has become more promising and attractive than before. We used the OPtimized Network Engineering Tool (OPNET) in our simulation studies [4]. OPNET follows the DES concept as the mean to analyze the system performance and behavior. OPNET can be described as a set of decision support tools and finite state machines. It has been designed to support the modelling and simulation of a large range of communication systems from a single LAN to a global satellite network. It is equipped with a range of tools, which allow developers to specify models in great detail, to identify elements of the model of interest, to execute the simulation, and to analyze generated output data.

A seven-cell model is adopted in the simulation, and a maximum of 12 channels per cell is allowed for mobile users. This assumption is valid under the very small cell size scenario. System contains 92 mobile users, with a call generating rate of 10 calls/hour per mobile user and an average call length of 20 minutes. At the center of each cell is a base station through which all mobile users transmit. All base stations are connected to a single device called MTSO (Mobile Telephone Switching Office). At any instance, each mobile user is controlled by a specific base station and logically belongs to that cell. When a mobile user leaves a cell, its base station notices the decrease in its signal strength. The base station then asks all surrounding base stations how much power they are getting from it. The current base station will transfer ownership to the cell getting the strongest signal. The mobile is then informed of its new base station and assigned a new channel.

We compare the traffic under different scenarios as shown in Table 1. Our goal is to investigate effects of two mechanisms: "priority handoff" and "differentiated QoS service class" on system utilization and handoff call blocking rates. Test sets (1) and (2) consist of audio signals without any differentiated QoS classes. The system utilization and handoff dropping rate in two test sets are compared to illustrate the effect of the use of priority handoff mechanism in test set (2). In test sets (3) and (4), we include audio and video, and each media type has an equal number of users. In test set (4), audio and video are allocated to the high and low priority classes, respectively. The simulation is run for 60 minutes for each test set.

The Effect on System Utilization and Handoff Dropping Rate with Priority Handoff

System utilization for test sets (1) and (2) is shown in Fig. 6.10(a). In test set (2), the system treats handoff calls with a higher priority than new calls, whereas no such priority scheme is used in test set (1) (i.e. the handoff as well as new calls are treated with the same priority). The system utilization is a little lower in test set (2) than in test set (1) since the system reserves some resources for higher priority handoff calls. However, as observed in Fig. 6.10(b), the handoff dropping rate for test set (2) is greatly reduced compared with that for test set (1).

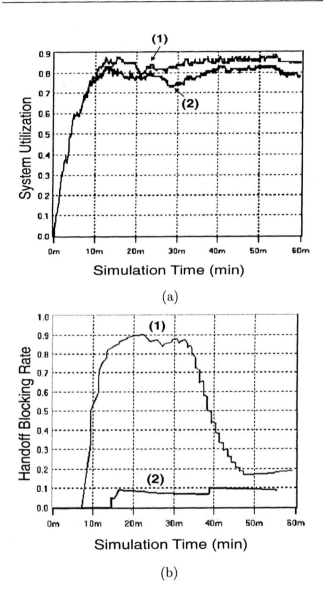

Figure 6.10: (a) System utilization without (test set 1) and with (test set 2) priority handoff mechanism. (b) Handoff dropping rate without (test set 1) and with (test set 2) priority handoff mechanism.

Table 6.6: Traffic Test Sets

	Test Set 1	*Test Set 2*	*Test Set 3*	*Test Set 4*
Traffic	100% Audio	100% Audio	50% Audio 50% Video	50% Audio 50% Video
Bandwidth (units)	1 (Audio)	1 (Audio)	1 (Audio) 2 (Video)	1 (Audio) 2 (Video)
Preferential treatments	NO	New v.s. Handoff	NO NO	New v.s. Handoff
QoS Priority classes	NO	NO	NO	2 classes

The Effect on System Utilization and Handoff Dropping Rate with Priority Handoff and Differentiated QoS Classes

System utilization and handoff dropping rates are compared for test sets (3) and (4) in Figs. 6.11(a) and 6.11(b), respectively. Both test sets (3) and (4) consist of audio and video data with an equal number of users. The priority handoff mechanism as well as the QoS service class are applied to data in test set (4), while none of them is used for test set (3). The system utilization is lower in test set (4) than that in test set (3) because the system reserves resources not only for handoff calls but also for the higher priority service class. However, as shown in Fig. 6.11(b), the handoff dropping rate for test set (4) is greatly reduced compared with that for test set (3).

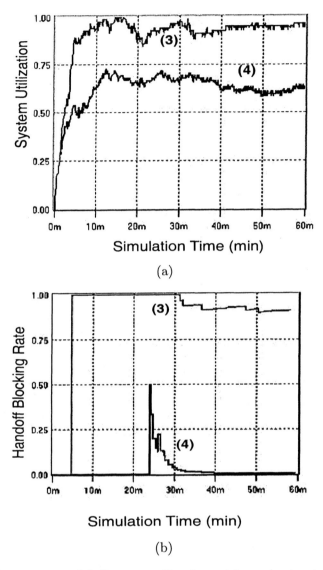

(a)

(b)

Figure 6.11: (a) System utilization without (test set 3) and with (test set 4) priority handoff and QoS classes. (b) Handoff dropping rate without (test set 3) and with (test set 4) priority handoff and QoS classes.

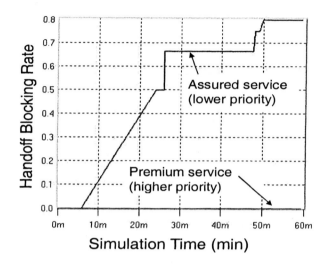

Figure 6.12: Handoff dropping rates

Comparison of the Handoff Dropping Rate between High and Low Priority Service Classes

Fig. 6.12 shows the handoff dropping rate comparison between the premium and the assured service QoS classes. We see that the handoff dropping rate is much lower for handoff calls in the premium service class. This is due to the fact that more resources have been reserved for handoff calls belonging to the premium class as compared to those belonging to the assured service class. As shown in Figs.6.11(b) and 6.12, our scheme can provide a lower blocking rate compared with the system without the use of any priority handoff mechanism and differentiated QoS class. Moreover, our scheme provides a lower blocking rate to the premium user.

6.3.2   Simulation Results for Dynamic Guard Channel Scheme -Scenario I

System and Service Model Parameters

Simulations are conducted by using OPNET [4] simulator. We have implemented our service model and CAC algorithm by using OPNET and compared the traffic under different scenarios. Our goal is to investigate the QoS measures in terms of system reward function defined in Eq. (6.19). We compare our proposed dynamic CAC scheme with the performance of fixed guard channel scheme.

In our simulation, we used a simple network topology with 7 cells as shown in Fig. (1.1), which covers a region in a non-overlapped fashion. Each cell has its own base station and cell capacity of 50 unit resources. A central control node, mobile switching center (MSC) is connected with each base station via a wired link. There are a number of mobile users with their own application profile in each cell, which can move across two or more cells according to their predetermined trajectories. Along its trajectory, a mobile user can originate connection requests randomly at its call generation rate. We assume a Poisson generation rate of connection requests, i.e. the inter-arrival time between two consecutive requests from a mobile user is exponentially distributed, and the connection duration is exponentially distributed. They are controlled by the following two parameters:

- λ: the mean request arrival rate in number of connections per hour

- l: the mean duration of each flow in minutes.

Parameter l is assigned the value of 20 minutes for each call connection. Increasing the value of λ results in the increment of network traffic load. We performed experiments for light traffic ($\lambda = 3$ and $l = 20$) and heavy traffic ($\lambda = 10$ and $l = 20$) conditions. We also apply a comprehensive service model, which includes properties of a call connection in its application profile:

- Φ_{max}, Φ_{min}: the maximum and minimum bandwidth requirement of a connection. Therefore, the bandwidth usage Φ is limited in the range of (Φ_{max}, Φ_{min}). In our simulation $\Phi \in \{1, 2, 4\}$.

- I_{rate}: the rate adaptivity indicator. $I_{rate} \in \{1, 0\}$.

- Π: the priority of a connection. $\Pi \in \{1, 2, 4, 8\}$ with equal probability in all simulations.

- M: the mobility of a connection. $\mu \in \{HIGH, LOW\}$.

Performance Comparison under Traffic with Multi-level Priority but without Rate Adaptive Characteristics (without Mobility Differentiation)

To evaluate the performance of the proposed scheme, we used the QoS performance in terms of system reward function, as defined in Eq. (6.19). The goal of this experiment is to evaluate the performance of proposed scheme in the presence traffic with multi-level priority classes under the traffic environment where all calls have no rate adaptive capability and mobility differentiation. Common traffic parameters used are $\Phi = \Phi_{max}$, $I_{rate}{=}0$, $\Pi \in \{1, 2, 4, 8\}$, and $\mu{=}LOW$. The following four schemes are compared:

- **S1**: proposed dynamic CAC and RRE scheme.

- **S2**: 0% fixed guard channel scheme.

- **S3**: 10% fixed guard channel scheme.

- **S4**: 20% fixed guard channel scheme.

Fig. 6.13 shows that our proposed dynamic scheme (S1) has best QoS performance in receiving the maximum reward among all the four schemes in light as well as heavy traffic load conditions.

Performance Comparison under Traffic with Multi-level Priority Classes and Rate Adaptive Characteristics (without Mobility Differentiation

The goal of this experiment is to evaluate the performance of proposed scheme in the presence of traffic with multi-level priority classes under the traffic environment where all calls are rate adaptive. Common traffic parameters are $\Theta = \{ \Theta_{min}, \Theta_{max} \}$, $I_{rate}{=}1$, $\Pi \in \{1, 2, 4, 8\}$, and $M{=}LOW$. The same four schemes are compared as in previous experiment. Fig. 6.14 shows that our proposed dynamic scheme (S1) has best QoS performance in receiving the maximum reward amongst the four schemes in light as well as heavy traffic load conditions.

Performance Comparison using Time-aware Weighting Factor in the Presence of Mobility Differentiation

To investigate the advantages of the use of time-aware weighting factor described in Eq. (6.13), the following two schemes, **Scheme A** and **Scheme B** are compared. Common traffic parameters used are $\Pi \in \{1, 2, 4, 8\}$, $M = \{LOW(75\%), HIGH(25\%) \}$. The bandwidth usage parameter used are $\Phi = \Phi_{max}$ and $\Phi = \{\Phi_{max}, \Phi_{min}\}$ in the presence of fixed-rate and adaptive rate traffic classes, respectively.

- **Scheme A**: The network support mobility and priority differentiation.

- **Scheme B**: The network uses different priority classes, but does not support mobility differentiation.

Figs. 6.15 and 6.16 show the simulation results for both schemes with fixed-rate and adaptive rate traffic classes (under light and heavy traffic load), respectively. Results show that dynamic Scheme A performs better due to the use of time-aware weighting factor in the resource reservation estimation(under light as well as heavy traffic load conditions).

6.3.3   Simulation Results for Dynamic Guard Channel Scheme -Scenario II

System and Service Model Parameters

A seven cells system is used in our simulation. Mobile terminals move in the system according to a certain trajectory and calls are generated in each MT, following the Poisson distribution. The call holding time is an exponential distribution. Cell capacity is 60 unit bandwidth for each cell and the maximum bandwidth requirement for each call supporting multimedia is 6 unit bandwidth. Traffic is classified into four priority classes, $Priority_1$ to $Priority_4$ where $Priority_1$ has highest priority level and class $Priority_4$ has lowest priority. We have assumed that each priority class has equal number of calls in the system. If calls are rate adaptive, they can be serviced in degraded quality mode at any discrete rate within the range of (Φ_{max}, Φ_{min}). If calls are not rate adaptive, they can only be serviced at a full rate. The average call holding time (l) is 20 minutes. Three mobility types (Low, Moderate and High speed) are equally distributed in the system.

Performance Comparison between Proposed Dynamic Scheme and Fixed GC Scheme

The performances of proposed dynamic CAC scheme and fixed GC schemes (0% GC, 5% GC) are compared for average call holding time $l = 20$ (min) and average call generating rate $\lambda = 10$ (calls/min/mobile). Results for rate adaptive and non-rate adaptive cases are illustrated in Figs. 6.17(a) and (b), respectively. The result shows that our proposed dynamic scheme out-performs 0% and 5% GC schemes in terms of global system reward R, in both rate adaptive as well as non-rate adaptive system.

Performance Comparison for Rate Adaptive and Non-rate Adaptive Systems

The QoS metrics in terms of handoff dropping rate $(P_{handoff})$ for each priority class are compared using proposed dynamic CAC scheme for $l = 20$ (min) and $\lambda = 10$ (calls/min/mobile). Results for rate adaptive and non-rate adaptive cases are shown in Fig. 6.18(a) and (b), respectively. The result shows that higher priority class will receive lower handoff dropping rate due to the use of resource reservation.

Performance Comparison among Different Priority Classes in Proposed Service Model using Dynamic Scheme

The performances between rate adaptive system and non-rate adaptive system are compared under different call generating rate λ ($= 3$, 6 and 10). QoS metrics in terms of reward R and system utilization are illustrated in Figs. 6.19(a) and 6.19(b), respectively. The results show that in rate adaptive system, both system reward and utilization will increase because calls are allowed to be serviced in a degraded mode when system is congested.

6.3.4 Conclusion and Future Work

In this chapter, we present a dynamic CAC and associated RR schemes based on the concept of guard channel to adapt the resource access priority by the use of SNR and distance information of the potential higher-priority calls in the neighboring cells, which are likely to handoff. Under light as well as heavy traffic conditions, our proposed CAC scheme out performs the fixed GC schemes. The cases for different application profiles of mobile terminals under various traffic conditions are also discussed.

We have considered a comprehensive service model, which includes not only mobile terminals' bandwidth requirements but also their different levels of priority, rate adaptivity, as well as their mobility. Our RR scheme provides more accurate estimation of the potential higher-priority calls arrival, and thus increases the system reward while providing QoS guarantees to higher-priority calls. Higher system reward implies that our proposed scheme can get a good balance between resource sharing and resource reservation to achieve the opposing goals of accommodating more calls, while providing QoS guarantees for high-priority classes connections.

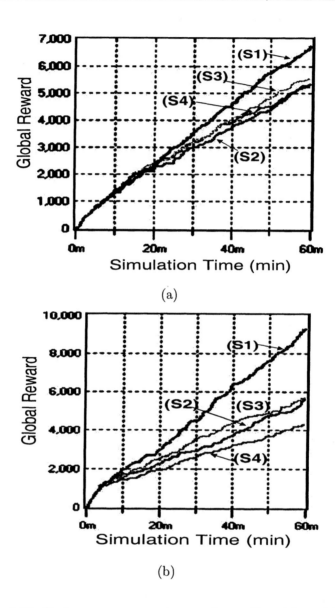

(a)

(b)

Figure 6.13: Performance comparison under traffic with multi-level priority but without rate adaptive characteristics (without mobility differentiation) for different schemes under (a) light traffic loading ($\lambda = 3$) and (b) heavy traffic loading ($\lambda = 10$).

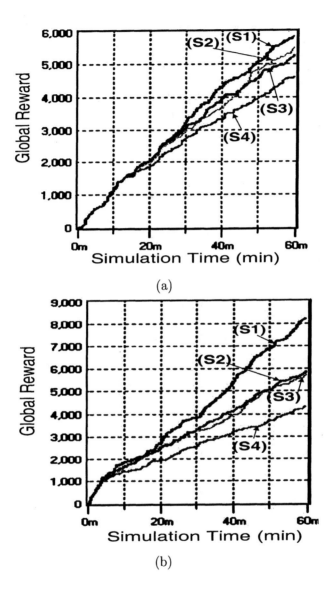

Figure 6.14: Performance comparison under traffic with multi-level priority classes and rate adaptive characteristics, without mobility differentiation for different schemes with rate adaptive abilities for all mobiles, under (a) light traffic loading ($\lambda = 3$) and (b) heavy traffic loading ($\lambda = 10$).

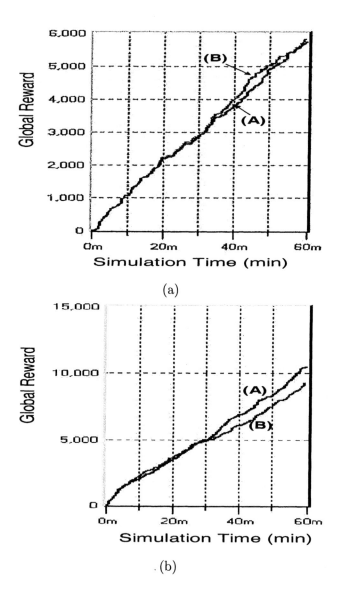

(a)

(b)

Figure 6.15: The performance comparison by using and not using time-aware weighting factor, in the presence of mobility differentiation with fixed-rate traffic class for all mobiles, under (a) light traffic loading ($\lambda = 3$) and (b) heavy traffic loading ($\lambda = 10$).

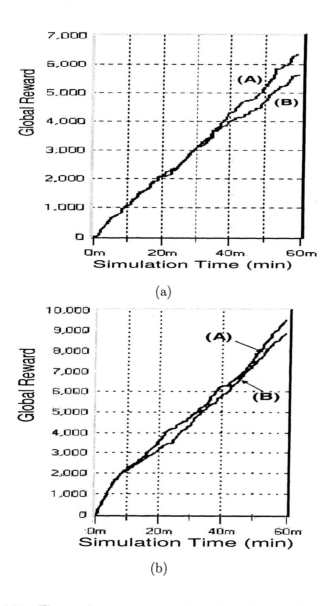

(a)

(b)

Figure 6.16: The performance comparison by using and not using time-aware weighting factor, in the presence of mobility differentiation with rate adaptive traffic class for all mobiles, under (a) light traffic loading ($\lambda = 3$) and (b) heavy traffic loading ($\lambda = 10$).

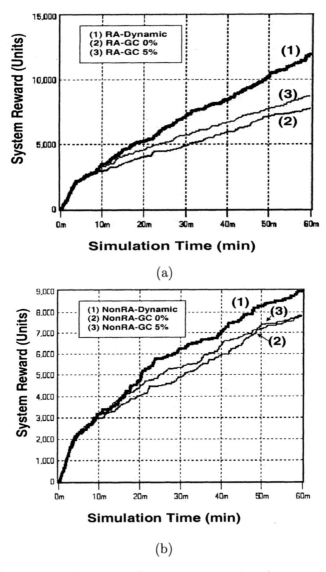

Figure 6.17: System performance comparison between proposed dynamic scheme and fixed GC scheme in (a) rate adaptive and (b) non-rate adaptive system.

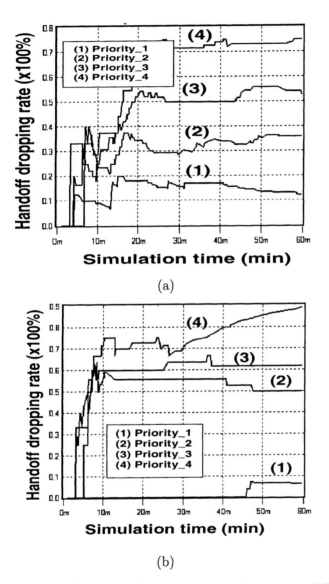

(a)

(b)

Figure 6.18: System performance comparison among different priority classes in the (a) rate adaptive and (b) non-rate adaptive system.

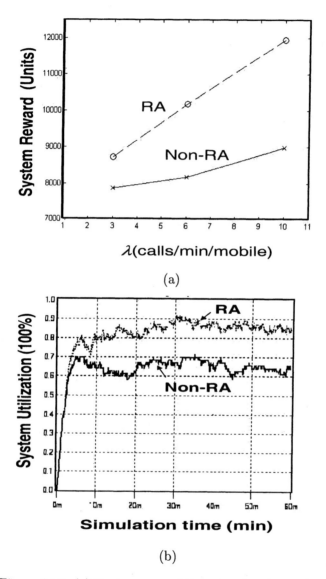

(a)

(b)

Figure 6.19: (a) System reward for rate adaptive and non-rate adaptive system, (b) System utilization for rate adaptive and non-rate adaptive system under traffic condition $\lambda = 10$, $l = 20$.

BIBLIOGRAPHY

[1] D. Hong and S. S. Rapport, "Traffic model and performance analysis for cellular mobile radiotelephone systems with prioritized and nonprioritized handoff procedures," *IEEE Transactions on Vehicle Technology*, vol. 35, pp. 77–92, Aug. 1986.

[2] Y. B. Lin, "Performance modeling for mobile telephone networks," *IEEE Network*, vol. 45, pp. 63–68, Nov.-Dec. 1997.

[3] W. C. Y. Lee, *Mobile Cellular Telecommunications - Analog and Digital Systems*, McGraw-Hill, 2000.

[4] I. Karzela, *Modeling and simulating communication networks: a hands-on approach using OPNET*, Prentice Hall, New Jersey, Aug. 1998.

Chapter 7

HANDOFF SCHEMES IN CDMA SYSTEMS

7.1 Introduction

The third generation (3G) wireless systems target at broadband wireless multimedia services. The Wide-band CDMA (W-CDMA) technology has emerged as the main air interface for 3G wireless systems, which promises to provide a transmission rate from 144Kbps to 2Mbps, enabling multimedia services as those provided by broadband wired networks [1, 2]. To meet the large bandwidth requirement of multimedia traffic, it is important to utilize the system resources efficiently and provide preferential treatment according to mobile user's application profile when the system is congested. The RRM in cellular network systems is responsible for efficiently utilizing air interface resources and guarantee a certain QoS level to different users according to their application profiles. The CAC mechanism is an important component of RRM. It affects the resource management efficiency and QoS guarantees provided to users.

In 2G TDMA/FDMA mobile systems, network accessibility controlled by the RRM module is typically designed based on the number of available channels. Due to the limited channel capacity, preferential treatment should be given to high priority calls to support them with higher QoS guarantees when the system is congested. Since dropping an ongoing call during handoff is less tolerable than blocking a new call, a handoff call should get a higher priority than a new call request. One way to provide preferential treatment is to pre-reserve a certain number of guard channels for higher priority calls such as handoff calls. Various GC schemes have been intensively

studied for 2G TDMA/FDMA wireless systems [3, 4, 5, 6, 7, 8]. Recently, dynamic GC schemes have been discussed in the literature to
improve system utilization while providing QoS guarantees to higher
priority calls [5, 6, 7, 8]. However, the GC approach is not completely suitable for CDMA systems because their capacity is limited
by the maximum tolerable interference in the system instead of available channels. In other words, a new call request in CDMA systems
is admitted if it does not introduce excessive interference into the
system.

Knutsson *et al.* [9] investigated the CAC scheme for downlink
communication for CDMA systems. Due to the asymmetric traffic
conditions in the reverse link (from the mobile to the base-station)
and the forward link (from the base-station to the mobile), the CAC
scheme should admit a call only when the call admission requirements are met in both directions. However, the reverse link capacity is usually more constrained in CDMA systems [10, 11, 12] and
should receive more attention. Huang and Yates [10] and Dimitriou
and Tafazolli [11] presented CAC schemes based on transmission
power. Liu and El Zarki [12] proposed a signal-to-interference ratio
(SIR)-based CAC scheme for the reverse link in DS-CDMA systems
to improve the system performance under heavy traffic conditions.
They assumed that the base station received the same signal power
from each of its mobile users, and CAC was designed based on the
variation of the SIR value. However, such an assumption does not
hold in practical systems, where power control is used to keep SIR
close to a target value during the whole operation for each mobile
user according to link conditions [13]. Shin *et al.* [14] proposed an
interference-based channel assignment scheme for DS-CDMA Cellular Systems. However, their CAC algorithm was based on fixed
resource reservation, where a fixed number of channels is reserved to
give the preferential treatment to high priority handoff calls.

In this chapter, we present a dynamic resource management
scheme for multimedia traffic. A RRE function is implemented in
each base station to assist RRM module to dynamically adjust the
level of resource needed to be reserved for the use of higher priority
traffic. The RRE function in each base station estimates the amount
of resource in terms of IGM by considering traffic load in its current

cell as well as traffic conditions in neighboring cells. The system resource is allocated efficiently by using dynamic resource reservation estimation and rate-adaptive CAC. In our proposed scheme, a constant target Signal to Interference Ratio (SIR) value is assumed due to the use of power control in practical systems.

The total interference level in the system is computed by employing the load curve introduced by Holma and Laakso [15]. The use of the load curve makes it possible to handle different levels of interference-increase introduced by heterogeneous traffic with various service rates. The salient features of our proposed scheme are summarized below. First, proposed scheme supports rate adaptive characteristics for multiple services with flexible QoS guarantees. Second, it takes heterogeneous traffic mobilities into consideration to achieve better resource estimation. Third, by using adaptive RRE, the amount of reserved resource can be dynamically changed by referencing the traffic condition in neighboring cells. Fourth, the proposed scheme bridges two important concepts, the GC and the load curve, that we call the interference guard margin (IGM) . Our contribution also lies in the fact that IGM is dynamically adjusted to fit the surrounding traffic conditions to maximize the objective function. The resulting CAC scheme gives preferential treatment to higher priority handoff calls by pre-reserving a certain amount of IGM.

The rest of this chapter is organized as follows. We provide an overview of capacity and load estimation for CDMA systems in Section 7.2. In Section 7.3, the interference guard margin (IGM) scheme to provide preferential treatment to mobile users in CDMA systems is proposed. It includes the CAC scheme and the associated dynamic RRE method. Several QoS metrics are measured in terms of the cost function, the handoff dropping probability, and the new call blocking probability. Section 7.4 shows the simulation results conducted with OPNET by using a service model. Finally, concluding remarks and future work are given in Section 7.5.

7.2   Overview of Capacity and Load Estimation in CDMA Systems

The capacity of a CDMA system is limited by the total interference it can tolerate, which is why it is called the interference-limit system. In CDMA systems, each new mobile user contributes to the overall level of interference, and call blocking occurs when the overall interference level reaches some level above background noise [16]. Normally, the interference level increases rapidly when the system load reaches a certain level. Users with different application profiles and attributes, such as the service rate, the signal-to-Interference ratio (SIR) requirement, etc., introduce a different amount of interference to the system. These factors are especially important in 3G cellular networks that support multimedia services. Liu and El Zarki [12], Holma and Toskala [15], and Viterbi $et.al.$ [16] have studied the effect of interference increase for traffics with the service rate R_i and target SIR requirement $\epsilon_i \equiv E_b/N_0$ for user i. ϵ_i is determined by the QoS requirement such as the bit error rate (BER) for a specific media type. In the proposed IGM scheme, we follow the similar mathematical treatment to derive the amount of interference introduced by user i with data rate R_i and ϵ_i.

In this chapter, we consider the resource management for uplink of W-CDMA system with chip-rate of $W = 3.84 Mcps$. The target ϵ_i is specified for a user i carrying traffic with data rate R_i. The processing gain of user i is denoted as G_i, which can be written as $G_i = \frac{W}{R_i}$. We then define P_N and S_i as the background noise power and the received power from user i, respectively, at the base station. Under these conditions, the target SIR ϵ_i can be written as:

$$\epsilon_i \equiv (\frac{E_b}{N_0})_i = \frac{G_i \cdot S_i}{\sum_{j=1, j \neq i}^{N} S_j + P_N} \tag{7.1}$$

where E_b is the energy per user bit, and N_0 is the (background) noise plus interference spectral density.

Let I_{total} denote the total received power at the base station from

N active users in the cell, we obtain

$$I_{total} = \sum_{i=1}^{N} S_i + P_N \qquad (7.2)$$

Here, the interference from own cell is considered. If inter-cell interference effect is considered, the "other cell interference factor" f has to be taken into account. The standard value of f is 0.55 in IS-95 as described in [2, 17]. I_{total} is the maximum planned power, which can be determined by the maximum planned load as well. The value of I_{total} is restricted to be smaller than the upper-bound I_{Th}. Otherwise the system becomes unstable and the overall interference increases dramatically.

Rewriting Eq. (7.1) by plugging the value of I_{total}, we get:

$$S_i = (1 + \frac{G_i}{\epsilon_i})^{-1} \cdot I_{total} \equiv \rho_i \cdot I_{total}, \qquad (7.3)$$

where

$$\rho_i = (1 + \frac{G_i}{\epsilon_i})^{-1} = \frac{\epsilon_i}{\epsilon_i + G_i} \qquad (7.4)$$

is called the load factor for user i [15, 2]. The total system load factor ρ is defined as the sum of load factors from N active mobile users, i.e., $\rho = \sum_{i=1}^{N} \rho_i$. From Eq. (7.1) and Eq. (7.2), we can write:

$$I_{total} - P_N = \sum_{i=1}^{N} S_i = \sum_{i=1}^{N} \rho_i \cdot I_{total} = \rho \cdot I_{total}. \qquad (7.5)$$

Shapira and Padovani [18] and Holma et al. [15, 2] estimated the interference increase by taking into account the load curve as shown in Fig. 7.1. They further defined the noise-rise, η, as the ratio of I_{total} to background noise P_N. The noise-rise η in Fig. 7.1 can be written as

$$\eta \equiv \frac{I_{total}}{P_N} = \frac{\sum_{i=1}^{N} S_i + P_N}{P_N} = (1 - \rho)^{-1}. \qquad (7.6)$$

By taking the partial derivative of I_{total} with respect to ρ, Holma [15] derived the relationship between I_i and ρ_i as below.

$$I_i = \frac{\rho_i}{1-\rho} \cdot I_{total} = \frac{(1 + \frac{G_i}{\epsilon_i})^{-1}}{1-\rho} \cdot I_{total} = \frac{(1 + \frac{W}{\epsilon_i \cdot R_i})^{-1}}{1-\rho} \cdot I_{total}. \quad (7.7)$$

This equation implies that when a mobile user i is admitted into the cell, the total interference increases by the amount of I_i, which can be expressed in term of its data rate R_i and target ϵ_i.

7.3   Proposed IGM Scheme

In this section, we develop an efficient radio resource management scheme based on the concept of IGM to provide preferential treatment to higher priority calls.

7.3.1   Service Model

In a mobile communication system with N active mobile users, the ith $(i < N)$ user's application profile, which characterizes its services, is described as

$$\Im(i) = \{r_i, (R_{max}, R_{min})_i, \Pi_i, M_i\}, \qquad (7.8)$$

where r_i, $(R_{max}, R_{min})_i$, Π_i and M_i in $\Im(i)$, denote user i's rate adaptivity, service rate range, priority, and mobility, respectively. The proposed service model is designed to take advantage of modern coding schemes and advanced mobile communication technologies as described below.

First, r_i is a binary indicator that indicates whether the user can be serviced at reduced bit-rates when the system is congested. To maintain a specified QoS level, a wireless system has to adapt to varying traffic conditions. Our proposed call admission control scheme can achieve this goal by exploiting the rate adaptive features (please see next paragraph) of modern multimedia coding schemes. Second, the service rate range $(R_{max}, R_{min})_i$ describes the target bandwidth consumption. If the network has enough resources, the request can be admitted at R_{max}. If the cellular system is overloaded (congested), a rate-adaptive user can be serviced at a lower rate (down to $\frac{R_{max,i}}{2}$ or even R_{min}) with degraded quality of service. Adaptation only takes place at the time of admitting new calls or at handoff epochs. Third, the priority tag Π_i helps the system to identify high priority users, who are likely to receive better QoS guarantees. Finally, three mobility types, M_i, are considered in our service model (high, moderate, and low mobility). The speed for high, moderate and low mobility traffic are 1, 2 and 4 unit speeds. Each different mobility traffic has a different weighting factor to estimate the amount of resources necessary to be reserved. This is discussed in our proposed resource reservation estimator in Section 7.3.3.

7.3.2 Resource Estimation and Reservation in IGM Scheme

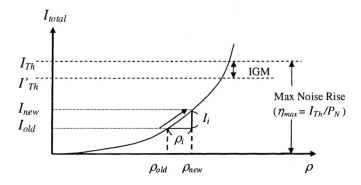

Figure 7.1: The load curve and the load estimation.

IGM is a natural extension of the guard channel idea developed in the context of TDMA/FDMA systems by considering the load factor for system capacity estimation in CDMA systems. As illustrated in Fig. 7.1, we have the following two operations. First, the load curve is used to estimate the load increase as well as the interference increase. Second, a certain amount of IGM, instead of guard channels, is pre-reserved for high priority calls. The amount of IGM is dynamically adjusted by the RRE function.

For a new call to be admitted, the total interference level should not exceed the upper bound of the interference with threshold I_{Th} that the system can tolerate. In addition to the constraint of I_{Th}, a lower priority call should comply with the augmented constraint I'_{Th}. The margin between I_{Th} and I'_{Th} is exactly the guard margin, which provides the preferential treatment to high priority calls by limiting the access to the low priority calls.

7.3.3   Dynamic Resource-Reservation Estimation

When a mobile terminal moves away from the BS toward its cell boundary, some of the neighboring base stations (BS) will receive a stronger signal from it. The mobile terminal will likely handoff to one of these cells. In our simulation system, the information about these neighboring cells is recorded in a handoff candidate registration (HCR) table that is assumed to be maintained at the corresponding MSC.

The HCR table provides useful information for estimating the mobile terminals that are likely to handoff to a given cell from its neighboring cells. We use this information to estimate the amount of resource in terms of interference margin $IGM(j)$ needed to be reserved when admitting a low priority call j in the cell. Call j will be admitted into the cell only when the resulting net interference of the system is less than $I_{Th} - IGM(j)$ after admitting this call.

The distance d_i for user i is measured based on the received signal power at the target base station. The radio propagation model is designed based on the analysis given in [22]. It is however difficult to measure distance d_i correctly based on the radio propagation model alone due to varying effects of shadowing, fading, etc.. The Global Positioning System (GPS) technology is becoming popular at a fast pace. The recent E-911 ruling issued by Federal Communications Commission (FCC) requires that the cellular operators must be able to accurately locate mobile callers requesting emergency services via 911. We can thus assume that cellular systems will soon be equipped with the technology that will accurately measure the distance d_i between the BS and mobile terminal. When the positioning technology becomes more advanced, accurate estimation of user velocity will also become possible. Chiu and Bassiouni [23] have recently proposed a scheme that predicts the handoff requests based on mobile positioning.

The resource in terms of IGM needed to be reserved for incoming call j is estimated based on the weighted sum of estimated minimum interference-increments, $I_{min,i}$, according to the application profile, for each potential handoff call from neighboring cells as given by:

$$
\begin{aligned}
IGM(j) &= \alpha \cdot \sum_{i \in S(j)} \omega_i \cdot I_{min,i} \qquad\qquad (7.9) \\
&= \alpha \cdot \sum_{i \in S(j)} \omega_i \cdot \left(\frac{\rho_{min,i}}{1 - \rho} \cdot I_{total} \right) \\
&= \alpha \cdot \sum_{i \in S(j)} \omega_i \cdot \frac{(1 + \frac{W}{\epsilon_i \cdot R_{min,i}})^{-1}}{1 - \rho} \cdot I_{total}
\end{aligned}
$$

where α, $0 \le \alpha \le 1$, is an empirical scaling factor that takes into account the fact that either some calls i from neighboring cells which are likely to handoff in the current cell terminate before they actually arrive or ongoing calls in the current cell terminate (or handoff to other cells). The results for IGM by using $\alpha = 1$ and $\alpha = 0.7$ are compared in Section 7.4.

In Eq. (7.9), $S(j)$ is a set consisting of all neighboring active calls who are moving toward current cell and whose priority is higher than current call request. An example is given later in this section to illustrate how the set $S(j)$ is determined. Furthermore, the weighting factor ω_i for user i in (7.9) is proportional to the ratio of mobility M_i to user's distance d_i from the base station, i.e.

$$
\omega_i \propto (M_i/d_i) \equiv T_i^{-1},
$$

where T_i is represented in unit of time. The factor ω_i implies that a high speed mobile user could be more likely to handoff into the current cell even when it is farther from the base station as compared to a low speed mobile user. Let us define

$$
\omega_i = \begin{cases} T_{Th}/T_i, & \text{if } T_i > T_{Th}, \\ 1, & \text{if } T_i < T_{Th}, \end{cases} \tag{7.10}
$$

where the threshold T_{Th} is an empirical value (in time unit) determined by a critical distance d_{Th} of a call from the target cell boundary with a typical mobile speed of 30 miles per hour. For instance, if we use a value of $d_{Th} = 400$ meter and a mobile speed of 30 miles/hr (800 m/min), the value of T_{Th} is approximately 0.5 min.

Therefore, any mobile user whose estimated arrival time T_i is less than 0.5 min is very likely to handoff into the current cell and we use $\omega_i = 1$, to reserve minimum resources requested by it. For T_i greater than T_{Th}, we reserve the resources partially. For another example, if T_i is estimated as 1.0 min, we use $\omega_i = 0.5$ and reserve one half of the requested minimum resources.

The distance d_i for user i is measured based on the received signal power at the target base station. The radio propagation model is designed based on the analysis done in [22].

Next, we show an efficient way to determine set $S(j)$ in Eq. (7.9). This set consists of all neighboring active calls that meet two criteria. First, the priority $\Pi(i)$ of call i is higher than that of incoming call j denoted by $\Pi(j)$. Second, the target cell $\Lambda^*(j)$ of call j is in the set of the handoff candidate cells $\Lambda(i)$ of call i in the HCR table. Here, the target cell $\Lambda^*(j)$ is defined as the neighboring base station with maximum received signal power among the handoff candidate base stations for call j.

Note also that the current cell of call i is not the same as the target cell of incoming call j. To conclude, we have

$$
S(j) = \{i|\Pi(i) > \Pi(j), \Lambda^*(j) \in \Lambda(i)\}. \tag{7.11}
$$

Figs. 7.2 (a) and (b) illustrate criteria $\Pi(i) > \Pi(j)$ and $\Lambda^*(j) \in \Lambda(i)$

for set $S(j)$, respectively. In Fig. 7.2(a), we show an incoming call j that requests a handoff from its current base station BS_3 toward its target cell BS_0. We will find out neighboring calls w.r.t. BS_0 whose priority is higher than that of call j. From this operation, we get $\{i|\Pi(i) > \Pi(j)\} = \{i_2, i_3, i_5\}$. In Fig. 7.2 (b), the target cell BS_0 is chosen as the best candidate cell from j's HCR table. Let us denote cell BS_0 by $\Lambda(j)^*$. In this figure, we see several dotted lines associated to each call, which represent handoff candidate cells for each call i. For example, i_2, i_3 and i_5 calls have $\{BS_5\}$, $\{BS_0, BS_6\}$ and $\{BS_0, BS_2\}$, respectively, as their candidate cells. Here, only calls i_3 and i_5 satisfy the condition $\{\Lambda^*(j) \in \Lambda(i)\}$. From the results of (a) and (b) in Eq. (7.11), we get $S(j) = \{i_3, i_5\}$.

7.3.4   Call Admission Control Algorithm

The pseudo code of the CAC algorithm for a media type with three scalable rates is given in Fig. 7.3, where a new call or a handoff call can be admitted into the system with three data rates : R_{max}, R_{half} and R_{min}. It can be generalized to a media type consisting of even more rates. Note that IGM_{new} and $IGM_{handoff}$ are the estimated bandwidths required to be reserved for new and handoff calls, respectively.

The basic concept behind CAC is to test whether there is enough system resource left to serve the current call request at a certain rate, after reserving the necessary resource for preferential treatment. The CAC test is performed according to the following steps:

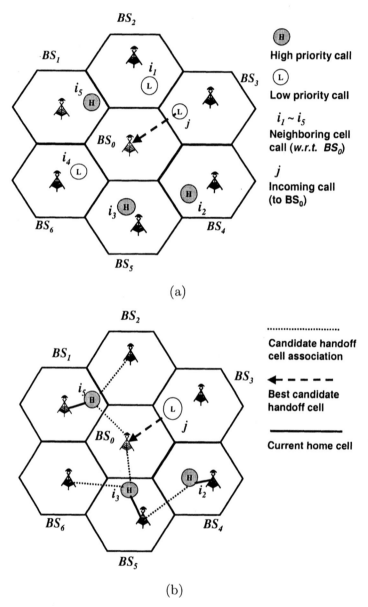

(a)

(b)

Figure 7.2: Set $S(j)$ in resource-reservation estimation: (a) $\Pi(i) >$ $\Pi(j)$ and (b)$\Lambda(j)^* \in \Lambda(i)$.

```
01    If INCOMING CALLS ARE NEW CALLS
02        If CALLS ARE NON-RATE ADAPTIVE
03            If (I_current + △I_i) < (I_Th − IGM_new)
04                ADMIT CALL REQUEST WITH RATE R_i
05            Else
06                REJECT CALL REQUEST
07        Else /*CALLS ARE RATE ADAPTIVE*/
08            If (I_current + △I_max,i) < (I_Th − IGM_new)
09                ADMIT CALL REQUEST WITH RATE R_max,i
10            Else If(I_current + △I_half,i) < (I_Th − IGM_new)
11                ADMIT CALL REQUEST WITH RATE R_half,i
12            Else If(I_current + △I_min,i) < (I_Th − IGM_new)
13                ADMIT CALL REQUEST WITH RATE R_min,i
14            Else
15                REJECT CALL REQUEST
16    Else /*INCOMING CALLS ARE HANDOFF CALLS*/
17        If CALLS ARE NON-RATE ADAPTIVE
18            If (I_current + △I_i) < (I_Th − IGM_handoff)
19                ADMIT CALL REQUEST WITH RATE R_i
20            Else
21                REJECT CALL REQUEST
22        Else /*CALLS ARE RATE ADAPTIVE*/
23            If (I_current + △I_max,i) < (I_Th − IGM_handoff)
24                ADMIT CALL REQUEST WITH RATE R_max,i
25            Else If(I_current + △I_half,i) < (I_Th − IGM_handoff)
26                ADMIT CALL REQUEST WITH RATE R_half,i
27            Else If(I_current + △I_min,i) < (I_Th − IGM_handoff)
28                ADMIT CALL REQUEST WITH RATE R_min,i
29            Else
30                REJECT CALL REQUEST
```

Figure 7.3: The proposed call admission control algorithm.

Step 1. *New or handoff call test*: An incoming call is first identified as a new or a handoff call type to decide its priority.

Step 2. *Rate adaptivity test*: The rate adaptivity of a new call (handoff call) is tested to decide whether it can be serviced at a lower data rate if the system is congested.

Step 3. *Non-rate adaptive call test*: If the call is rate-adaptive, go to *Step 4*. Otherwise, test whether the amount of interference after admitting the current call and reserving the estimated IGM will exceed the maximum interference level that the system can tolerate.

Step 4. *Rate adaptive call test*: If the call is rate-adaptive, the current call could be serviced at rates of $R_{max,i}$, $R_{half,i}$, and $R_{min,i}$, depending on the system traffic condition. The amount of interferences introduced by a call are $\triangle I_{max,i}$, $\triangle I_{half,i}$, and $\triangle I_{min,i}$ when it is serviced at rates $R_{max,i}$, $R_{half,i}$, and $R_{min,i}$, respectively. Then, we test the admission criteria by the order of data rates for the highest to the lowest. The call is served at its highest admissible rate.

7.4   Simulation Results

7.4.1   System Model and Link Characteristics

Simulations were conducted by using the OPNET [24]. The link characteristics of the CDMA system used in simulation are given below. A network topology with seven cells is used. Normally, the noise rise would be 2 to 4 [2]. When system become heavily loaded, e.g. ρ close to 0.9, the noise rise could rise up to 10 as shown in Fig. 8.2 in [2]. An empirical value of the maximum interference level I_{Th} is then set to ten times of background noise, e.g. $\eta_{max} = 10$ for simulation purpose. The same radio frequency band is reused for every cell, and separated frequency bands are used for the reverse link and the forward link. There are 420 mobile terminals with three types of mobility (equally distributed). Each cell has its own home base station. Several neighboring base stations together are connected with a centralized center such as the MSC (or RNC) via a wired link. There are a number of mobile users with their own application profiles in each cell, which can move across two or more cells according to their predetermined trajectories. Along its trajectory, a mobile user can originate connection requests randomly at its call generation rate.

The Poisson call arrival rate and the exponentially distributed call holding time are assumed. The call arrival rate and the call holding time are controlled by two parameters, i.e. λ (the mean request arrival rate measured in the number of connections per hour) and l (the mean call holding time of each flow in minutes, which is set to 15 for each call connection). Increasing the value of λ results in the increment of the network traffic load.

Values used in the application profile $\Im(i)$ of user i are listed below:

1) $r_i \in \{YES, NO\}$,
2) $R_{max,i}$ are, respectively, set to 19.2 Kbps, 38.4 Kbps and 76.8 Kbps for voice, audio and video traffics and $R_{min,i}$ is set to be $\frac{R_{max,i}}{2}$,
3) $\Pi \in \{new, handoff\}$,
4) $M_i \in \{HIGH, MOD, LOW\}$,
5) Communication system parameters used in simulation include: CDMA chip rate $W = 3.84 Mcps$ and target SIR $\varepsilon_i = 7dB$.

For simulation purpose, we define a weighted cost function J as below for performance comparison when preferential treatment is employed.

$$
\begin{aligned}
J &= w_n \cdot P_n + w_h \cdot P_h \\
&= w_n \cdot \frac{N_{\text{new\_block}}}{N_{\text{new\_request}}} + w_h \cdot \frac{N_{\text{handoff\_block}}}{N_{\text{handoff\_request}}},
\end{aligned}
\tag{7.12}
$$

where $N_{\text{new\_block}}$ is the total number of new calls blocked, $N_{\text{new\_request}}$ the total number of new calls requested, $N_{\text{handoff\_block}}$ the total number of handoff calls blocked, and $N_{\text{handoff\_request}}$ the total number of handoff calls requested. Given a new-call blocking weighting $\omega_n = 1$, we used the handoff-call dropping weighting $\omega_h = 10$ for J to reflect the higher cost for dropping a handoff call.

To illustrate the advantage of dynamic IGM , we compare the QoS performance in terms of the cost function J given in Eq. (7.12) for the following four scenarios:

- Non-priority scheme (also referred to as the complete sharing scheme),
- Fixed IGM 20% scheme (i.e. IGM is fixed to 20% I_{Th}),
- Dynamic IGM scheme (with resource-reservation scaling factor $\alpha = 1$)

7.4.2   Non-rate Adaptive Traffic

Figs. 7.4(a) and (b) show the performance under light to heavy traffic load with λ varying from 0.1 to 1.3 (calls per hour per user). Fig. 7.4(a) shows that the dynamic IGM scheme has the best QoS performance (evaluated in terms of cost function J). Fig. 7.4(b) shows that the dynamic IGM scheme significantly reduces the handoff dropping probability P_h without much increase in the new call blocking probability P_n as compared to the non-priority scheme.

7.4.3   Rate Adaptive Traffic

The rate adaptive traffic can be admitted into a system with a lower data rate when the system is congested. Figs. 7.5(a) and (b) show the system performance under light to heavy traffic load with λ varying from 0.1 to 1.3. The performance comparison in terms of the cost function J is given in Fig. 7.5(a). The fixed scheme cannot adapt to traffic condition and possibly reserve excessive resource, this will lead to a higher new-call blocking probability P_n. Results show that the performance in terms of J for a fixed scheme is not necessarily better than that of the non-priority scheme. However, the dynamic IGM scheme can adapt well to each traffic condition, and has the best QoS performance. The new call blocking probability P_n and the handoff call dropping probability P_h are shown in Fig. 7.5(b).

Again, the dynamic IGM scheme has the best performance under the light as well as the heavy traffic loads.

We compare the cost function J for non-rate adaptive as well as rate adaptive cases in Fig. 7.6. It is clear that the proposed dynamic IGM scheme outperforms the non-priority scheme for both rate adaptive and non-rate adaptive cases. Furthermore, the dynamic IGM scheme achieves a more significant improvement when the rate adaptive mechanism is available. Note that, in order to address the issues of some calls from the neighboring cells will not handoff to the current cell and some other ongoing calls in the current cell will either terminate or handoff to other cells. Thus, a scaling factor α is used to avoid excessive resource reservation in the choice of IGM.

7.4.4 System Utilization

Let system utilization U be defined as a ration of the time average of the occupied resource to the total resource ratio, i.e., $U = \mathbf{E}[\frac{BW_{occu}}{BW_{total}}]$, where BW_{occu} and BW_{total} are instant occupied resource and total system resource, respectively. $\mathbf{E}[\cdot]$ denotes time average over 5 hour simulation time.

System utilizations for all four schemes w.r.t. non-rate adaptive and rate-adaptive cases is compared in Fig. 7.7 under very heavy traffic load (i.e. $\lambda = 5$). We see that system utilization is higher for traffic with the rate-adaptive capability than that without the rate-adaptive capability. This can be explained by the fact that the system can provide calls with a reduced data rate when the system is congested, thus increasing the overall system utilization.

Due to resource reservation adopted by the proposed dynamic IGM scheme, it cannot fully utilize the system resource in order to provide preferential treatment to higher priority calls. The use of scaling factor $\alpha = 0.7$ in IGM increases the system utilization for non-rate adaptive traffic at the expense of dropping more handoff calls.

As shown in the figure, the non-priority scheme has the best system utilization performance since it does not reserve any resources for handoff calls, and accepts the calls on the first-come first-served basis. The fixed IGM scheme gives higher system utilization as compared to the dynamic IGM scheme because the latter reserves more resources to serve handoff calls at the heavy load condition.

Finally, we would like to point out that system utilization is about the same for the non-priority scheme, the fixed IGM and the dynamic IGM under the light traffic condition. System utilization is also not sensitive to the rate-adaptive capability of the underlying traffic. Usually, the maximum rate demanded by a user can be served.

7.5 Conclusion and Future Work

Effective radio resource management schemes, including dynamic RRE and CAC, based on the concept of IGM for CDMA systems were presented. We considered a service model that included mobile terminals' service rate, their different levels of priority, rate adaptivity, as well as their mobility. The proposed dynamic IGM scheme reserves a certain amount of interference margin for high priority handoff calls by referencing the traffic condition and mobile users' application profile in neighboring cells. The mobility-aware weighted sum plays an important role in the RRE process so that the effect of different mobility is taken into consideration. It was shown by computer simulation that the proposed fixed and dynamic IGM schemes outperform the non-priority scheme in giving a smaller cost function J under light as well as heavy traffic conditions.

In modern CDMA systems, soft handoff is employed to provide a better transition process than hard handoff. The benefit of soft-handoff is that a mobile user can connect two or more base stations at the same time, thus greatly reducing the probability of call-dropping due to severe channel impairments. However, soft-handoff should be used only up to a certain extent because an excessive amount of soft-handoff connections increases the downlink interference. In our future work, the IGM scheme will be generalized to CDMA systems with hybrid hard- and soft-handoff schemes to achieve an efficient radio resource management mechanism.

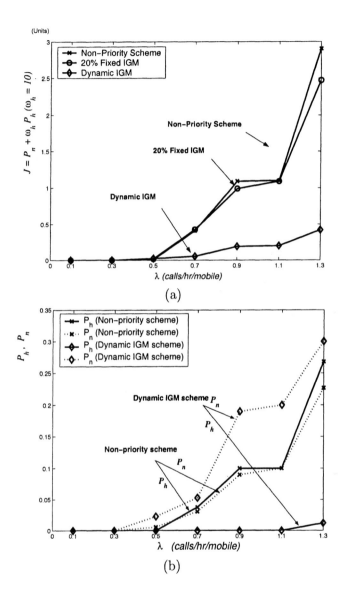

Figure 7.4: Performance comparison for non-rate adaptive users under light to heavy traffic load: (a) the cost function J and (b) the new call blocking rate P_n and the handoff dropping rate P_h.

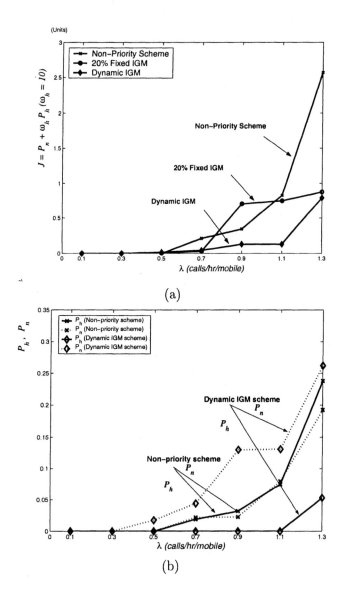

Figure 7.5: Performance comparison for rate adaptive users under light to heavy traffic load: (a) the cost function J and (b) the new call blocking rate P_n and the handoff dropping rate P_h.

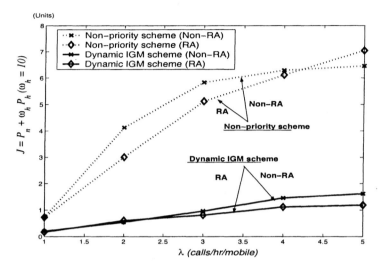

Figure 7.6: Performance comparison between rate adaptive (RA) and non-rate adaptive (Non-RA) schemes for the cost function J.

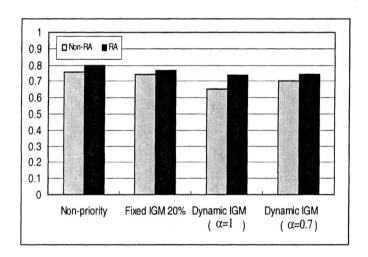

Figure 7.7: Comparison of system utilization w.r.t. rate-adaptive and non-rate-adaptive traffic under the heavy load with $\lambda = 5$.

BIBLIOGRAPHY

[1] J. Zander, S.-L. Kim, M. Almgren, and O. Queseth, *Radio Resource Management for Wireless Networks*, Artech House Publishers, 2001.

[2] H. Holma and A. Toskala, *WCDMA for UMTS, Radio Access For Third Generation Mobile Communications*, John Wiley & Sons, June 2000.

[3] D. Hong and S. S. Rapport, "Traffic model and performance analysis for cellular mobile radiotelephone systems with prioritized and nonprioritized handoff procedures," *IEEE Transactions on Vehicle Technology*, vol. 35, pp. 77–92, Aug. 1986.

[4] H. Chen, S. Kumar, and C.-C. J. Kuo, "Differentiated QoS aware priority handoff in cell-based multimedia wireless network," in *IS&T/SPIE's 12th Int. Symposium, Electronic Imaging 2000*, vol. 3974, (San Jose, CA), pp. 940–948, Jan. 2000.

[5] M. Naghshineh and M. Schwartz, "Distributed call admission control in mobile/wireless networks," *IEEE Journal Selected Areas Communications*, vol. 14, pp. 711–717, May 1996.

[6] P. Ramanathan, K. M. Sivalingam, P. Agrawal, and S. Kishore, "Dynamic resource allocation schemes during handoff for mobile multimedia wireless networks," *IEEE Journal on Selected Areas in Communications*, vol. 17, pp. 1270–1283, July 1999.

[7] A. S. Acampora and M. Naghshineh, "Control and quality of service provisioning in high-speed micro-cellular networks," *IEEE Personal Communications*, No. 2nd Quarter, pp. 36–43, 1994.

[8] A. Sutivong and J. M. Peha, "Novel heuristics for call admission control in cellular systems," *1997 IEEE 6th International Conference on Universal Personal Communications Record*, vol. 1, pp. 129–133, 1997.

[9] J. Knutsson, P. Butovitsch, M. Persson, and R. D. Yates, "Downlink admission control strategies for CDMA systems in a Manhattan environment," in *Proc. VTC*, vol. 2, pp. 1453–1457, May 1998.

[10] C. Huang and R. Yates, "Call admission in power controlled CDMA systems," in *Proc. VTC*, vol. 3, pp. 1665 – 1669, May 1996.

[11] N. Dimitriou and R. Tafazolli, "Resource management issues for UMTS," in *Proc. 1st IEEE Int. Conf. 3G Mobile Commun. Technol.*, pp. 401 – 405, March 2000.

[12] Z. Liu and M. E. Zarki, "SIR-based call admission control for DS-CDMA cellular systems," *IEEE Journal on Selected Areas in Communications*, vol. 12, pp. 638–644, May 1994.

[13] K. Gilhousen, R. P. I.M. Jacobs, and et. al, "On the capacity of a cellular CDMA system," *IEEE Transactions On Vehicle Technology*, vol. 40, pp. 303–312, 1991.

[14] S. M. Shin, C.-H. Cho, and D. K. Sung, "Interference-based channel assignment for DS-CDMA cellular systems," *IEEE Transactions On Vehicle Technology*, vol. 48, pp. 233–239, Jan. 1999.

[15] H. Holma and J. Laakso, "Uplink admission control and soft capacity with mud in CDMA," in *Proceedings of VTC'99 Fall*, (Amsterdam, Netherlands), pp. 19–22, Sept. 1999.

[16] A. M. Viterbi and A. J. Viterbi, "Erlang capacity of a power controlled CDMA system," *IEEE Journal on Selected Areas in Communication*, vol. 11, No. 6, pp. 892–900, 1993.

[17] Vijay K. Garg, *IS-95 CDMA and CDMA2000*, Prentice Hall, 2000.

[18] J. Shapira and R. Padovani, "Microcell engineering in CDMA cellular networks," *IEEE Trans. on Vehicular Technology*, vol. 43, pp. 213–216, 1994.

[19] I. JTC1/SC29/WG11, "Generic coding of moving pictures and associated audio information," *ISO/IEC International Standard 13818*, Nov. 1994.

[20] I. JTC1/SC29/WG11, "Overview of the MPEG-4 standard," *ISO/IEC N3747*, Oct. 2000.

[21] I. JTC1/SC29/WG1, "JPEG2000 part 1 final committee draft version 1.0," *ISO/IEC International Standard N1646R*, Mar. 2000.

[22] W. C. Y. Lee, *Mobile Cellular Telecommunications - Analog and Digital Systems*. McGraw-Hill, 2000.

[23] M. -H. Chiu and M. A. Bassiouni, "Predictive schemes for hand-off prioritization in cellular networks based on mobile positioning," *IEEE J. Sel. Areas in Commun.*, vol. 18, 2000, pp. 510–522.

[24] I. Karzela, *Modeling and simulating communication networks: a hands-on approach using OPNET*, Prentice Hall, NJ, Aug. 1998.

Chapter 8

RADIO RESOURCE MANAGEMENT WITH MDP

8.1 Introduction to MDP and CAC Policy

The GC scheme was proved by Ramjee *et al.* [6] to be optimal for an objective function formed by a linear weighting of the new call blocking and the handoff dropping probabilities. However, the GC scheme is not optimal for multiple traffic classes. In order to achieve better QoS, it is necessary to have a stochastic control policy. One well known stochastic control approach is the Markov Decision Process (MDP), which is a branch from operation research. MDP models are widely used in diverse research areas and practical applications such as ecology [7, 8], economics [9], and network routing [10]. Extensive application and examples are given in [11, 12]. In this chapter, we focus on the application of MDP to the design of the call admission control policy.

Stochastic control with dynamic programming is often used in determining the optimal call admission control policy. Rezaiifar, Makowski, and Kumar [13] developed a dynamic programming algorithm to design the optimal handoff strategy in cellular radio systems. They studied the problem of choosing the optimal fixed threshold that minimizes the cost function associated with switching the cell site and maximizes the reward for improving the quality of the call. While dynamic programming can be used to solve simplified problems as given in [14], the high computational cost as a result of the large state space size makes dynamic programming practically infeasible.

Unlike dynamic programming, MDP can be used to derive an optimal call admission policy in a stationary sense. In most cases, where the traffic condition does not change rapidly, the MDP-based call admission policy provides a better trade-off between optimality and complexity. In the literature, MDP was used to determine the optimal call admission policy for priority treatment. Two types of calls, new call and hard-handoff traffic, were considered by Ho [15] to maximize system utilization in deriving the optimal call admission policy. In this work, the accepting preference was given to hand-off calls by limiting the dropping rate of handoff calls. Choi *et al.* [16] studied the highway traffic control system with multiple traffic classes to maximize the revenue. Xiao, Chen, and Wang [17] applied the model to rate-adaptive multimedia traffic to re-allocate the system resource to different media types to maximize the revenue.

The Semi-Markov Decision Process (SMDP) generalizes the Markov Decision Process (MDP) in the following characteristics [11].

- (1) SMDP allows the decision maker to choose actions whenever the system state changes.

- (2) SMDP Models the system evolution in continuous time. (MDP also has a continuous-time version.)

- (3) SMDP allows the time spent in a particular state to follow an arbitrary probability distribution.

According to the description above, the MDP can be viewed as a special case of a SMDP in which inter-transition times are exponentially distributed and actions are chosen at every transition. More details can be found on pp. 530 in Puterman's book [11].

In this chapter, we formulate the optimal RRM design as the MDP optimization problem. The stationary optimal CAC policy, which can be used in the hybrid handoff scenario (i.e. hard and soft handoff schemes are both) is determined by solving a set of linear programming equations. The CAC policy can be controlled by choosing appropriate actions to accept or reject calls of different classes according to the current system state.

In this work, we examine the relationship between optimal policy and traffic parameters. To be more specific, we address the preferential treatment problem by assigning high priority calls with a larger weighting factor. Experimental results show that CAC can be derived to achieve the best result in terms of weighted system utilization by using the MDP approach. Our work can be easily extended to a rich service model with multiple traffic types and models. Several distinguishing features of the proposed scheme are highlighted below.

- The MDP model is used for system modeling while the linear programming (LP) technique is adopted to find the optimal CAC policy in our work. The MDP model was used to determine the optimal CAC policy for new call and hard-handoff traffic [15] as well as rate-adaptive traffic [17] before. However, previous work did not address the behavior of the optimal CAC policy. The CAC policy will be studied for two types of calls in Section 8.2. Unlike previous work, we focus on the relationship between the optimal CAC policy and traffic parameters.

- We propose an advanced CAC policy, which is used in the hybrid handoff scenario (with both hard-handoff and soft-handoff traffic coming from neighboring cells and modelled by MDP). In the CDMA system, hard-handoff is applied when the target cell and its adjacent cells operate at a different frequency. This is referred to as inter-frequency handoff . Soft handoff is adopted when the target cell and its adjacent cells operate at the same frequency, which is referred to as intra-frequency handoff . If there are multiple active CDMA carrier frequencies, independent frequency synthesizers would be required in soft handoff. This is however costly, and hard handoff is chosen in such a scenario as well.

- The proposed scheme is suitable for traditional channel-limit systems as well as interference-limit CDMA systems using system capacity estimation as described in Section 8.1.1. Thus, both systems can be handled within the same framework under one CAC policy.

- Soft handoff is implemented in modern CDMA systems to take advantage of macro-diversity to enhance signal quality.

However, more system resources are required since soft handoff calls connect two or more base stations simultaneously during handoffs.

- From the perspective of CAC precision, the system state of the multiple threshold guard channel scheme as described in [4] can be viewed as a subset of the system state model proposed in this chapter. The proposed optimization scheme has a higher controlling precision (or a higher system state resolution) and it outperforms the complete sharing and the multiple threshold guard channel schemes.

The rest of this chapter is organized as follows. First, system capacity estimation used in the proposed model is discussed for interference-limit systems in Section 8.1.1. With capacity estimation , we are able to measure how much resource to allocate, how to conduct the system plan, and how to perform the optimal CAC policy accordingly. The MDP formulation is introduced in Section 8.1.2. It is followed by the LP solution in Section 8.1.5. In Sections 8.2 and 8.3, optimal CAC policies are derived using the MDP formulation for homogeneous and hybrid handoff systems, respectively, and numerical results are given and discussed for both cases. In Section 8.5, the complexity of the MDP approach for 2-dimension and 3-dimension cases is analyzed. Finally, concluding remarks and future work are described in Section 8.5.3.

8.1.1   System Capacity in Interference-limited Systems

Most optimal CAC schemes have been developed for the second generation (2G) cellular systems, in which system resources are divided into "channels" in the unit of time slots with time division multiple access (TDMA) or in the unit of non-overlapping narrow frequency bands in frequency division multiple access (FDMA). The system capacity in a code division multiple access (CDMA) system is however quite different. The capacity of the third generation (3G) CDMA system is limited by the overall interference the system can tolerate. Such a system is called the interference-limit system, in which each mobile user admitted to the system contributes a certain amount of interference to the overall interference. Normally, the interference level increases rapidly when the system load is beyond a threshold level. In other words, call blocking occurs when the overall interference level reaches the maximum level that the system can tolerate.

This behavior leads to two issues: (1) how to estimate the system capacity for the CDMA system and (2) how to design CAC policies for CDMA systems according to a finite system state rather than a continuous value of system load, which is proportional to the overall interference level. In the following, we briefly derive the system capacity in terms of the overall interference and the minimum service rate (or the data rate). The CAC policy will be designed based on the bandwidth consumption state, which belongs to a finite and countable set. Thus, the system load derivation simultaneously solves the above two issues.

It is assumed that the system can provide services to user i at data rate $R_i = n_i R$, where R is the minimum data rate and n_i is an integer. Let us focus on user 1, and let P_{total} be the total interference power experienced by user 1 from other $N-1$ active users, and P_N be the background noise power. We have

$$P_{total} = \sum_{i=2}^{N} S_i + P_N. \tag{8.1}$$

The maximum value of P_{total} is limited by an upper-bound that is determined by the system stability threshold η as defined in [18], [19]. ($\eta = 0.1$ in [18], [19]). The system is unstable and the overall interference increases dramatically when the ratio of the overall interference power to the background noise power exceeds the value of $\frac{1}{\eta}$. Therefore, for the system to be stable, we have the following constraint:

$$\frac{P_{total}}{P_N} \leq \frac{1}{\eta}. \tag{8.2}$$

As a result of (8.1) and (8.2), we have

$$P_{total} = \sum_{i=2}^{N} S_i + P_N = \sum_{i=2}^{N} n_i \cdot R \cdot (E_b)_i + P_N$$

$$\geq \sum_{i=2}^{N} n_i \cdot R \cdot (E_b)_i + P_{total} \cdot \eta, \tag{8.3}$$

where S_i is the received power at the base station from user i and E_b is the energy per user bit. Under the assumption of perfect power control, a fixed target energy to-interference-ratio [18] ϵ_i for user i can be achieved. The parameter ϵ_i is defined as $\epsilon_i \equiv (E_b/I_0)_i$, and $P_{total} = I_0 \cdot W$. Then, one can rewrite (8.3) to be

$$\sum_{i=1}^{N} n_i \cdot R \cdot (E_b)_i \approx \sum_{i=2}^{N} n_i \cdot R \cdot (E_b)_i \leq P_{total} \cdot (1 - \eta) = I_0 \cdot W(1 - \eta)$$

$$\sum_{i=1}^{N} n_i \cdot \epsilon_i \leq (W/R) \cdot (1 - \eta), \tag{8.4}$$

where I_0 is the interference spectral density and W is the chip rate for CDMA system. In the case of the same target value ϵ for all users, we can rewrite the above equation as

$$\sum_{i=1}^{N} n_i \leq \frac{(W/R) \cdot (1 - \eta)}{\epsilon}. \tag{8.5}$$

Eq. (8.5) indicates that the total capacity of a CDMA system should not exceed

$$C = \lfloor \frac{(W/R) \cdot (1 - \eta)}{\epsilon} \rfloor, \tag{8.6}$$

which can be viewed as the allowed maximum bandwidth of the system. We will use this system capacity estimation to design the CAC policy in the next Section.

8.1.2 MDP and Optimal CAC Policy

MDP is a dynamic process that can model an optimization problem in which the time intervals between consecutive decision epochs are not identical but follow a probability distribution. In our proposed model, the consecutive decision epochs are assumed to follow the exponential distribution.

Consider a dynamic process which is observed at discrete time points $t = 0, 1, 2, \dots$. At each observation, the system is classified to be one of the possible states, which are finite and countable. The set of possible states forms the state space denoted by X. For each state $\mathbf{x} \in \mathbf{X}$, a set of actions denoted by $A_\mathbf{x}$ is given, which is again finite and countable. Note that, for a continuous-time Markov decision process (CT-MDP), a standard uniformization [12] technique can be used to handle CT-MDP with the solution developed for the discrete time case (e.g. the linear programming technique).

It is worthwhile to point out that choosing a stationary CAC policy means to find a mapping from the state space to the action space. When a call event happens, we can determine the optimal action according to the current system state based on such a mapping. In this chapter, we discussed handoff scenarios in Section 8.2 and Section 8.3 for homogeneous and hybrid handoff systems, respectively.

8.1.3 Markov Decision Process Model

The proposed MDP model can be uniquely identified by the following five components: the decision epochs, the state space, the action space, the reward function, and the transition probabilities.

- Decision epochs
  The decision is made only at the occurrence of a call arrival. Call arrival events include new call and handoff call arrivals. At events of call termination or handoff to other cells, decisions will not be made.

- State space
  The state space X is a set of all possible combinations of occupied channels of k types in the system, i.e.

$$X = \left\{ \mathbf{x} | \mathbf{x} = (x_1, x_2, \cdots, x_k); x_1, \cdots, x_k \geq 0, \sum_{i=1}^{k} x_i \leq C \right\},$$

$$(8.7)$$

where x_i is the number of calls for call type i, and the maximum system capacity can be expressed as $C = \lfloor \frac{(W/R)(1-\eta)}{\varepsilon} \rfloor$.

- **A**ction space
  The action space A is a set of vectors consisting of k binary elements, *i.e.*

$$A = \{\mathbf{a}|\mathbf{a} = (a_1, a_2, \cdots, a_k); \tag{8.8}$$
$$a_1, a_2, \cdots, a_k \in \{0(\text{reject}), 1(\text{accept})\}\},$$

where a_i are actions for each type of calls, respectively. They take the value of 0 for rejecting and 1 for accepting that type of calls. The action space $A_\mathbf{x}$ for state $\mathbf{x} \in X$ can be written as

$$A_\mathbf{x} = \begin{cases} \{\mathbf{a} = (1, 1, \cdots, 1)\}, & \text{if } \mathbf{x} = (0, 0, \cdots, 0), \\ \{\mathbf{a} = (0, 0, \cdots, 0)\}, & \text{if } \sum_{i=1}^{k} x_i = C, \\ \{\mathbf{a}|\mathbf{a} = (a_1, a_2, \cdots, a_k); & \\ \quad a_1, a_2, \cdots, a_k \in \{0, 1\}\}, & \text{Otherwise.} \end{cases} \tag{8.9}$$

- **R**eward function
  Let us denote call arrival event as a vector $\mathbf{e}$ of k binary values,

$$\left\{ \mathbf{e} = (e_1, e_2, \cdots, e_i, \cdots, e_k) | e_i \in \{0, 1\}, \sum_{i=1}^{k} e_i = 1 \right\} \tag{8.10}$$

The above equation means that only one of the k binary values is equal to one, say $e_i = 1$. It indicates that the current arrival event is call type i. Other values of e_j, $j \in \{1, \cdots, k\}, j \neq i$, are set to 0. This notation will be convenient for the following reward definition.

The reward $r(\mathbf{x}, \mathbf{a})$ defined below is earned if the system state is in state $\mathbf{x}$, and the CAC policy is configured as $\mathbf{a}$:

$$r(\mathbf{x}, \mathbf{a}) = \sum_{i=1}^{k} w_i(x_i + e_i \cdot a_i), \tag{8.11}$$

where w_i's are weighting factors for all call types. When the weighting factors are equal to one, the objective reward function is to maximize system utilization.

- **The transition probabilities.**
  Let $\tau(\mathbf{x}, \mathbf{a})$ be the sojourn time, the expected time until a new state is entered [15, 17], when the system is in the present state $\mathbf{x} \in X$ when action $\mathbf{a} \in A_{\mathbf{x}}$ is chosen. The value of sojourn time can be expressed by

$$\tau(\mathbf{x}, \mathbf{a}) = \frac{1}{\sum_{i=1}^{k} \lambda_i a_i + \sum_{i=1}^{k} x_i \mu_i}, \qquad (8.12)$$

where a_i's represent actions determined from the optimal CAC policy for each call type. They take binary values, *i.e.* with 1 for accepting a call and 0 for rejecting a call.

The transition probability from state $\mathbf{x}$ with action $\mathbf{a}$ to state $\mathbf{y}$ can be written as

$$P(\mathbf{y}|\mathbf{x}, \mathbf{a}) = \begin{cases} a_1 \cdot \lambda_1 \cdot \tau(\mathbf{x}, \mathbf{a}), & \text{if } \mathbf{y} = \mathbf{x} + (1, 0, \cdots, 0), \\ a_2 \cdot \lambda_2 \cdot \tau(\mathbf{x}, \mathbf{a}), & \text{if } \mathbf{y} = \mathbf{x} + (0, 1, \cdots, 0), \\ \vdots & \vdots \\ a_k \cdot \lambda_k \cdot \tau(\mathbf{x}, \mathbf{a}), & \text{if } \mathbf{y} = \mathbf{x} + (0, 0, \cdots, 1), \\ x_1 \cdot \mu_1 \cdot \tau(\mathbf{x}, \mathbf{a}), & \text{if } \mathbf{y} = \mathbf{x} - (1, 0, \cdots, 0), \\ x_2 \cdot \mu_2 \cdot \tau(\mathbf{x}, \mathbf{a}), & \text{if } \mathbf{y} = \mathbf{x} - (0, 1, \cdots, 0), \\ \vdots & \vdots \\ x_k \cdot \mu_k \cdot \tau(\mathbf{x}, \mathbf{a}), & \text{if } \mathbf{y} = \mathbf{x} - (0, 0, \cdots, 1), \\ 0, & \text{if } \mathbf{y} = \mathbf{x}. \end{cases}$$

$$(8.13)$$

8.1.4   Uniformization Technique

In this section, we describe a uniformization technique [12], which transforms a continuous-time Markov chain with non-identical decision times, denoted by M, into an equivalent continuous-time Markov process, denoted by $\overline{M}$, in which decision epochs are generated by a Poisson process at a uniform rate. After the use of uniformization, the transition process from one state to another can be described by a discrete-time Markov chain which allows fictitious transitions from a state to itself, while process M does not allow transition back to the starting state.

By observing the transition probability defined in Eq. (8.13), we see that the state transition rate from state $\mathbf{x}$ to $\mathbf{y}$ is either λ or μ depending on the action at the transition epoch. The time spent in state $\mathbf{x}$ before transition to state $\mathbf{y}$ is $\tau(\mathbf{x}, \mathbf{a})$, if $\mathbf{x} \neq \mathbf{y}$. It is obvious that the time spent in each state is varying according to the value of $\tau(\mathbf{x}, \mathbf{a})$. To transform the non-identical mean of transition times into an equivalent continuous time Markov process, let us define

$$\tau_c = (\sum_{i=1}^{k} \lambda_i a_i + C \cdot \max\{\mu_1, \mu_2, \cdots, \mu_k\})^{-1}. \qquad (8.14)$$

Furthermore, let us define a continuous-time Markov process $\overline{M}$ with an identical state transition duration of mean τ_c with the following transition probability

$$\overline{P(y|x,a)} = \begin{cases} a_1 \cdot \lambda_1 \cdot \tau_c, & \text{if } \mathbf{y} = \mathbf{x} + (1, 0, \cdots, 0), \\ a_2 \cdot \lambda_2 \cdot \tau_c, & \text{if } \mathbf{y} = \mathbf{x} + (0, 1, \cdots, 0), \\ \vdots & \vdots \\ a_k \cdot \lambda_k \cdot \tau_c, & \text{if } \mathbf{y} = \mathbf{x} + (0, 0, \cdots, 1), \\ x_1 \cdot \mu_1 \cdot \tau_c, & \text{if } \mathbf{y} = \mathbf{x} - (1, 0, \cdots, 0), \\ x_2 \cdot \mu_2 \cdot \tau_c, & \text{if } \mathbf{y} = \mathbf{x} - (0, 1, \cdots, 0), \\ \vdots & \vdots \\ x_k \cdot \mu_k \cdot \tau_c, & \text{if } \mathbf{y} = \mathbf{x} - (0, 0, \cdots, 1), \\ 1 - \frac{\tau_c}{\tau(\mathbf{x},\mathbf{a})}, & \text{if } \mathbf{y} = \mathbf{x} \end{cases}$$

$$= \begin{cases} \frac{\tau_c}{\tau(\mathbf{x},\mathbf{a})} \cdot P(y|x,a), & \text{if } \mathbf{x} \neq \mathbf{y} \\ 1 - \frac{\tau_c}{\tau(\mathbf{x},\mathbf{a})}, & \text{if } \mathbf{x} = \mathbf{y} \end{cases} \qquad (8.15)$$

$$\equiv \begin{cases} \alpha(\mathbf{x}, \mathbf{a}) \cdot P(y|x,a), & \text{if } \mathbf{x} \neq \mathbf{y} \\ 1 - \alpha(\mathbf{x}, \mathbf{a}), & \text{if } \mathbf{x} = \mathbf{y} \end{cases} \qquad (8.16)$$

where $\alpha(x, a) = \tau_c / \tau(\mathbf{x}, \mathbf{a})$.

Note that the net transitional probability of $\overline{P(\mathbf{y}|\mathbf{x}, \mathbf{a})}$ from state $\mathbf{x}$ to $\mathbf{y}$, with $\mathbf{x} \neq \mathbf{y}$, is the same as $P(\mathbf{y}|\mathbf{x}, \mathbf{a})$ as shown below. For $\mathbf{x} \neq \mathbf{y}$, we have

$$\overline{P(\mathbf{y}|\mathbf{x},\mathbf{a})} = \sum_{i=0}^{\infty} (1 - \alpha(x, a))^i \cdot \alpha(x, a) \cdot P(\mathbf{y}|\mathbf{x}, \mathbf{a})$$

$$= \frac{\alpha(x,a)}{1-(1-\alpha(x,a))} \cdot P(\mathbf{y}|\mathbf{x},\mathbf{a}) \qquad (8.17)$$

$$= P(\mathbf{y}|\mathbf{x},\mathbf{a}). \qquad (8.18)$$

8.1.5   Solution via Linear Programming

The linear programming algorithm is a well known technique to find out Markov decision policies. It has several advantages. First, it is convenient to add more constraints without modifying the structure significantly. Second, it allows us to analyze the sensitivity of the obtained solution. The simplex method is commonly adopted to find the optimal solution. Instead of evaluating the objective function for all candidates satisfying the constraints, this method examines only "better" candidates, which are known in advance that the objective function will have a larger value [20]. If the decision process $\overline{M}$ is designed with an identical transition duration distribution, say $\tau_c = 1$, the linear programming algorithm can be used to find the maximum reward function. It is given below with decision variables $\pi(\mathbf{x},\mathbf{a})'$, $\mathbf{x} \in X$ and $\mathbf{a} \in A_{\mathbf{x}}$:

$$\text{Maximize} \sum_{\mathbf{x} \in X} \sum_{\mathbf{a} \in A_{\mathbf{x}}} r(\mathbf{x},\mathbf{a})\pi(\mathbf{x},\mathbf{a})' \qquad (8.19)$$

subject to

$$\sum_{\mathbf{a} \in A_y} \pi(\mathbf{y},\mathbf{a})' - \sum_{\mathbf{x} \in X} \sum_{\mathbf{a} \in A_x} \overline{P(\mathbf{y}|\mathbf{x},\mathbf{a})}\pi(\mathbf{x},\mathbf{a})' = 0, \forall\, \mathbf{y} \in X (8.20)$$

$$\sum_{\mathbf{x} \in X} \sum_{\mathbf{a} \in A_x} \pi(\mathbf{x},\mathbf{a})' = 1, \qquad (8.21)$$

$$\pi(\mathbf{x},\mathbf{a})' \geq 0, \quad \mathbf{x} \in X, \quad \mathbf{a} \in A_{\mathbf{x}}. \qquad (8.22)$$

For the discrete-time MDP case, the term $\pi(\mathbf{x},\mathbf{a})$ can be interpreted as the long term fraction of the decision epochs at which the system is in state $\mathbf{x}$ and action $\mathbf{a}$ is taken ([12], pp. 181).

By using the uniformization technique, which converts the transition probability structure from $P(\mathbf{y}|\mathbf{x},\mathbf{a})$ to probability structure $\overline{P(\mathbf{y}|\mathbf{x},\mathbf{a})}$, we can find that solving the standard Linear Programming formulation for the discrete-time MDP problem above is equivalent to that for an uniformized continuous-time MDP problem.

Let $\pi(\mathbf{x}, \mathbf{a}) = \pi(\mathbf{x}, \mathbf{a})'/\tau(\mathbf{x}, \mathbf{a})$, we can verify that (8.19) is equivalent to (8.24), and (8.21) is equivalent to (8.26). Furthermore, we will show below that (8.20) is equivalent to (8.25).

$$\sum_{\mathbf{a}\in A_y} \pi(\mathbf{y}, \mathbf{a})' - \sum_{\mathbf{x}\in X}\sum_{\mathbf{a}\in A_x} \overline{P(\mathbf{y}|\mathbf{x}, \mathbf{a})}\pi(\mathbf{x}, \mathbf{a})' = 0, \forall\, \mathbf{y} \in X,$$

$$\implies \sum_{\mathbf{a}\in A_y} \pi(\mathbf{y}, \mathbf{a})' - \sum_{\mathbf{a}\in A_x} \overline{P(\mathbf{x} = \mathbf{y}|\mathbf{x} = \mathbf{y}, \mathbf{a})}\pi(\mathbf{x}, \mathbf{a})'$$
$$- \sum_{\mathbf{x}\neq\mathbf{y},\mathbf{x},\mathbf{y}\in X}\sum_{\mathbf{a}\in A_x} \overline{P(\mathbf{y}|\mathbf{x}, \mathbf{a})}\pi(\mathbf{x}, \mathbf{a})' = 0, \forall\, \mathbf{y} \in X,$$

$$\implies \sum_{\mathbf{a}\in A_y} \pi(\mathbf{y}, \mathbf{a})'(1 - (1 - \alpha(x, a)))$$
$$- \sum_{\mathbf{x}\neq\mathbf{y},\mathbf{x},\mathbf{y}\in X}\sum_{\mathbf{a}\in A_x} \overline{P(\mathbf{y}|\mathbf{x}, \mathbf{a})}\pi(\mathbf{x}, \mathbf{a})' = 0, \forall\, \mathbf{y} \in X,$$

$$\implies \sum_{\mathbf{a}\in A_y} \pi(\mathbf{y}, \mathbf{a})'(\alpha(x, a)) - \sum_{\mathbf{x}\neq\mathbf{y},\mathbf{x},\mathbf{y}\in X}\sum_{\mathbf{a}\in A_x} \overline{P(\mathbf{y}|\mathbf{x}, \mathbf{a})}\pi(\mathbf{x}, \mathbf{a})' = 0, \forall\, \mathbf{y} \in X,$$

$$\implies \sum_{\mathbf{a}\in A_y} \frac{\pi(\mathbf{y}, \mathbf{a})'}{\tau_c}(\alpha(x, a)) - \sum_{\mathbf{x}\neq\mathbf{y},\mathbf{x},\mathbf{y}\in X}\sum_{\mathbf{a}\in A_x} \frac{\overline{P(\mathbf{y}|\mathbf{x}, \mathbf{a})}}{\frac{\tau_c}{\tau(\mathbf{x},\mathbf{a})}} \frac{\pi(\mathbf{x}, \mathbf{a})'}{\tau(\mathbf{x}, \mathbf{a})} = 0, \forall\, \mathbf{y} \in X,$$

$$\implies \sum_{\mathbf{a}\in A_y} \frac{\pi(\mathbf{y}, \mathbf{a})'}{\tau(\mathbf{x}, \mathbf{a})} - \sum_{\mathbf{x}\neq\mathbf{y},\mathbf{x},\mathbf{y}\in X}\sum_{\mathbf{a}\in A_x} \frac{\overline{P(\mathbf{y}|\mathbf{x}, \mathbf{a})}}{\alpha(x, a)} \frac{\pi(\mathbf{x}, \mathbf{a})'}{\tau(\mathbf{x})} = 0, \forall\, \mathbf{y} \in X,$$

$$\implies \sum_{\mathbf{a}\in A_y} \frac{\pi(\mathbf{y}, \mathbf{a})'}{\tau(\mathbf{x}, \mathbf{a})} - \sum_{\mathbf{x}\neq\mathbf{y},\mathbf{x},\mathbf{y}\in X}\sum_{\mathbf{a}\in A_x} \frac{\overline{P(\mathbf{y}|\mathbf{x}, \mathbf{a})}}{\alpha(x, a)} \frac{\pi(\mathbf{x}, \mathbf{a})'}{\tau(\mathbf{x}, \mathbf{a})} = 0, \forall\, \mathbf{y} \in X,$$

$$\implies \sum_{\mathbf{a}\in A_y} \frac{\pi(\mathbf{y}, \mathbf{a})'}{\tau(\mathbf{x}, \mathbf{a})} - \sum_{\mathbf{x},\mathbf{x},\mathbf{y}\in X}\sum_{\mathbf{a}\in A_x} P(\mathbf{y}|\mathbf{x}, \mathbf{a})\pi(\mathbf{x}, \mathbf{a}) = 0, \forall\, \mathbf{y} \in X,$$

$$\implies \sum_{\mathbf{a}\in A_y} \pi(\mathbf{y}, \mathbf{a}) - \sum_{\mathbf{x}\in X}\sum_{\mathbf{a}\in A_x} P(\mathbf{y}|\mathbf{x}, \mathbf{a})\pi(\mathbf{x}, \mathbf{a}) = 0, \forall\, \mathbf{y} \in X$$

$$(8.23)$$

where $\alpha(x, a) = \tau_c/\tau(\mathbf{x}, \mathbf{a})$ and $P(\mathbf{y}|\mathbf{x}, \mathbf{a}) = 0$, if $\mathbf{y} = \mathbf{x}$. We can therefore solve the following modified Linear Programming for the continuous-time MDP problem with the standard Linear Programming formulation for the discrete-time MDP problem after the uniformization transformation:

$$\text{Maximize} \sum_{\mathbf{x} \in X} \sum_{\mathbf{a} \in A_{\mathbf{x}}} \mathrm{r}(\mathbf{x}, \mathbf{a}) \tau(\mathbf{x}, \mathbf{a}) \pi(\mathbf{x}, \mathbf{a}) \tag{8.24}$$

subject to

$$\sum_{\mathbf{a} \in A_y} \pi(\mathbf{y}, \mathbf{a}) - \sum_{\mathbf{x} \in X} \sum_{\mathbf{a} \in A_x} P(\mathbf{y}|\mathbf{x}, \mathbf{a}) \pi(\mathbf{x}, \mathbf{a}) = 0, \forall \, \mathbf{y} \in X \tag{8.25}$$

$$\sum_{\mathbf{x} \in X} \sum_{\mathbf{a} \in A_x} \tau(\mathbf{x}, \mathbf{a}) \pi(\mathbf{x}, \mathbf{a}) = 1, \tag{8.26}$$

$$\pi(\mathbf{x}, \mathbf{a}) \geq 0, \quad \mathbf{x} \in X, \quad \mathbf{a} \in A_x. \tag{8.27}$$

The term $\tau(\mathbf{x}, \mathbf{a})\pi(\mathbf{x}, \mathbf{a})$ can be viewed as the long term fraction of the decision epochs at which the system is in state $\mathbf{x}$ and action $\mathbf{a}$ is taken [17].

The optimal action in each state can be chosen among all actions in each state since the value of $\pi(\mathbf{x}, \mathbf{a})$ will be zero for all but one action in each state. This implies that the optimal policy is non-randomized admission policy and the action $\mathbf{a}$ chosen for each state is a deterministic function of state $\mathbf{x}$.

Next, in Section 8.2, an optimal CAC policy used in a homogeneous system is modelled using MDP, and the behavior of the optimal CAC policy is discussed.

8.2   2-D MDP and Optimal CAC Policy in Homogeneous Handoff Systems

In this section, the CAC policy is studied for two types of calls. In contrast with previous work, we focus on the relationship between the optimal policy and traffic parameters.

8.2.1   System Model

Consider a cellular system with the maximum system capacity C as given in Eq. (8.6) for a CDMA system. Two types of calls are considered here: new call and handoff calls. In the homogeneous handoff system, only one type of handoff occurs. As described in [21], a soft handoff occurs while the mobile terminal is under the mobile station control and in the traffic channel state. This handoff is characterized by commencing communications with a new base station with the same CDMA frequency assignment before terminating communications with the old base station. A hard handoff occurs when the mobile terminal is transferred between disjoint active sets, where the CDMA frequency assignment changes, the frame offset changes, or the mobile station is directed from a CDMA traffic channel to an analog voice channel. Here, we focus on the hard handoff.

Consider that call arrivals follow the Poisson distribution with the following parameters (note that in a homogeneous handoff system, only a single type of handoff is present):

λ_n: the arrival rate of new calls
λ_h: the arrival rate of handoff calls
μ_n: the service rate of new calls
μ_h: the service rate of handoff calls

The MDP model for the proposed handoff scheme used in a homogeneous handoff system is stated below.

- Decision epochs correspond to time instances only at call arrivals.

- The state space X is defined as

$$X = \{\mathbf{x}|\mathbf{x} = (x_n, x_h), x_n \geq 0, x_h \geq 0, x_n + x_h \leq C\}, \quad (8.28)$$

where x_n and x_h are numbers of new calls and handoff calls, respectively, and $C = \lfloor \frac{(W/R)(1-\eta)}{\varepsilon} \rfloor$.

- The action space is a set of vectors consisting of two binary elements, i.e.

$$A = \{\mathbf{a}|\mathbf{a} = (a_n, a_h); a_n, a_h \in \{0(\text{reject}), 1(\text{accept})\}\}, \quad (8.29)$$

where a_n and a_h are actions for new calls and handoff calls, respectively. They take the value of 0 for rejecting and 1 for accepting that type of calls. The action space $A_\mathbf{x}$ for state $\mathbf{x} \in X$ can be written as

$$A_\mathbf{x} = \begin{cases} \{\mathbf{a} = (1,1)\}, & \text{if } \mathbf{x} = (0,0), \\ \{\mathbf{a} = (0,0)\}, & \text{if } x_n + x_h = C, \\ \{\mathbf{a}|\mathbf{a} = (a_n, a_h); a_n, a_h \in \{0,1\}\}, & \text{Otherwise.} \end{cases} \tag{8.30}$$

Let $\tau(\mathbf{x}, \mathbf{a})$ be the sojourn time in the present state $\mathbf{x} \in X$ when action $\mathbf{a} \in A_\mathbf{x}$ is chosen. We obtain

$$\tau(\mathbf{x}, \mathbf{a}) = \frac{1}{\lambda_n a_n + \lambda_h a_h + x_n \mu_n + x_h \mu_h}, \tag{8.31}$$

where a_n and a_h represent actions for new calls and handoff calls, respectively. They take a binary value, *i.e.* with 1 for accepting a call and 0 for rejecting a call.

- **Reward function**
  The call arrival event can be represented by a vector $\mathbf{e}$ consisting of two binary values,

$$e = (e_n, e_h) = \begin{cases} (1,0), & \text{if new call arrival,} \\ (0,1), & \text{if handoff call arrival} \end{cases} \tag{8.32}$$

The reward $r(\mathbf{x}, \mathbf{a})$, defined below, is earned if the system state is in state $\mathbf{x}$ and the CAC policy is configured as $\mathbf{a}$. The reward function can be written as

$$r(\mathbf{x}, \mathbf{a}) = w_n(x_n + e_n \cdot a_n) + w_h(x_h + e_h \cdot a_h), \tag{8.33}$$

where w_n and w_h are weighting factors for new and handoff calls, respectively.

- The transition probability from state $\mathbf{x}$ with action $\mathbf{a}$ to state $\mathbf{y}$ can be written as

$$P(\mathbf{y}|\mathbf{x}, \mathbf{a}) = \begin{cases} a_n \cdot \lambda_n \cdot \tau(\mathbf{x}, \mathbf{a}), & \text{if } \mathbf{y} = \mathbf{x} + (1,0), \\ a_h \cdot \lambda_h \cdot \tau(\mathbf{x}, \mathbf{a}), & \text{if } \mathbf{y} = \mathbf{x} + (0,1), \\ x_n \cdot \mu_n \cdot \tau(\mathbf{x}, \mathbf{a}), & \text{if } \mathbf{y} = \mathbf{x} - (1,0), \\ x_h \cdot \mu_h \cdot \tau(\mathbf{x}, \mathbf{a}), & \text{if } \mathbf{y} = \mathbf{x} - (0,1), \\ 0, & \text{if } \mathbf{y} = \mathbf{x}. \end{cases} \tag{8.34}$$

8.2.2   Modified Linear Programming

By using the uniformization technique, we can derive a modified linear programming algorithm associated with the MDP for the maximum reward function. It is given below with decision variables $\pi(\mathbf{x}, \mathbf{a})$, $\mathbf{x} \in X$ and $\mathbf{a} \in A_{\mathbf{x}}$.

$$\text{Maximize} \sum_{\mathbf{x} \in X} \sum_{\mathbf{a} \in A_{\mathbf{x}}} r(\mathbf{x}, \mathbf{a}) \tau(\mathbf{x}, \mathbf{a}) \pi(\mathbf{x}, \mathbf{a}) \qquad (8.35)$$

subject to

$$\sum_{\mathbf{a} \in A_y} \pi(\mathbf{y}, \mathbf{a}) - \sum_{\mathbf{x} \in X} \sum_{\mathbf{a} \in A_x} P(\mathbf{y}|\mathbf{x}, \mathbf{a}) \pi(\mathbf{x}, \mathbf{a}) = 0, \forall \; \mathbf{y} \in X \quad (8.36)$$

$$\sum_{\mathbf{x} \in X} \sum_{\mathbf{a} \in A_x} \tau(\mathbf{x}, \mathbf{a}) \pi(\mathbf{x}, \mathbf{a}) = 1, \qquad (8.37)$$

$$\pi(\mathbf{x}, \mathbf{a}) \geq 0, \quad \mathbf{x} \in X, \quad \mathbf{a} \in A_{\mathbf{x}}. \qquad (8.38)$$

The variables $\pi(\mathbf{x}, \mathbf{a})$ satisfying (8.36)-(8.38) can be viewed as the steady-state probabilities of being in state $\mathbf{x}$ and choosing action $\mathbf{a}$. The optimal action in each state can be chosen among all actions in each state since the value of $\pi(\mathbf{x}, \mathbf{a})$ will be zero for all but one action in each state. This implies that the optimal policy is non-randomized admission policy and the action $\mathbf{a}$ chosen for each state is a deterministic function of state $\mathbf{x}$.

8.3   3-D MDP and Optimal CAC Policy

8.3.1   System Model

In this Section, we propose a 3-D MDP model for the hybrid handoff system with three types of calls. They are new calls, hard-handoff calls, and soft-handoff calls. The system capacity for CDMA with a hybrid handoff system is defined in Eq. (8.6). New, hard-handoff and soft-handoff call arrivals follow the Poisson distribution with parameters:

λ_n :   the average arrival rate of new calls;

λ_s :   the average arrival rate of soft handoff calls;

λ_h :   the average arrival rate of hard handoff calls;

μ_n :   the average service rate of new calls;

μ_s: the average service rate of soft handoff calls;

μ_h: the average service rate of hard handoff calls.

The proposed MDP model is defined by the following five components: the decision epochs, the state space, the action space, the reward function, and the transition probabilities. The MDP model for the proposed hybrid handoff system is stated below.

- The decision is made only at the occurrence of a call arrival. Call arrival events include new call and handoff call arrivals.

- The state space X is a set of all possible combinations of occupied channels of each type in the system, $i.e.$

$$X = \{\mathbf{x}|\mathbf{x} = (x_n, x_s, x_h), x_n \geq 0, x_s \geq 0, x_h \geq 0, x_n + x_s + x_h \leq C\},$$
(8.39)

where x_n, x_s, x_h are numbers of new calls, soft handoff calls, and hard handoff calls, respectively, and $C = \lfloor \frac{(W/R)(1-\eta)}{\varepsilon} \rfloor$.

- The action space is a set of vectors consisting of three binary elements, $i.e.$

$$A \;=\; \{\mathbf{a}|\mathbf{a} = (a_n, a_s, a_h); \tag{8.40}$$
$$a_n, a_s, a_h \in \{0(\text{reject}), 1(\text{accept})\}\},$$

where a_n, a_s, a_h are actions for new calls, soft handoff calls, and hard handoff calls, respectively. They take the value of 0 for rejecting and 1 for accepting that type of calls. The action space $A_\mathbf{x}$ for state $\mathbf{x} \in X$ can be written as

$$A_\mathbf{x} = \begin{cases} \{\mathbf{a} = (1,1,1)\}, & \text{if } \mathbf{x} = (0,0,0), \\ \{\mathbf{a} = (0,0,0)\}, & \text{if } x_n + x_s + x_h \geq C, \\ \{\mathbf{a}|\mathbf{a} = (a_n, a_s, a_h); & \\ \quad a_n, a_s, a_h \in \{0,1\}\}, & \text{Otherwise.} \end{cases}$$
(8.41)

Let $\tau(\mathbf{x}, \mathbf{a})$ be the sojourn time in the present state $\mathbf{x} \in X$ when action $\mathbf{a} \in A_{\mathbf{x}}$ is chosen. The sojourn time can be expressed as

$$\tau(\mathbf{x}, \mathbf{a}) = \frac{1}{\lambda_n a_n + \lambda_s a_s + \lambda_h a_h + x_n \mu_n + x_s \mu_s + x_h \mu_h} \quad (8.42)$$

- The reward $r(\mathbf{x}, \mathbf{a})$ of state $\mathbf{x}$ when action $\mathbf{a}$ is taken is expressed as

$$r(\mathbf{x}, \mathbf{a}) = w_n(x_n + e_n \cdot a_n) + w_n(x_s + e_s \cdot a_s) + w_h(x_h + e_h \cdot a_h), \quad (8.43)$$

where w_n, w_s, and w_h are the weighting factors for each call type, respectively. When the weighting factors are equal to one, the objective reward function is to maximize system utilization. The call arrival event can be represented by a vector $\mathbf{e}$ consisting of two binary values,

$$e = (e_n, e_s, e_h) = \begin{cases} (1, 0, 0), & \text{if new call arrival} \\ (0, 1, 0), & \text{if soft handoff call arrival} \\ (0, 0, 1), & \text{if hard handoff call arrival} \end{cases} \quad (8.44)$$

- The transition probability from state $\mathbf{x}$ with action $\mathbf{a}$ to state $\mathbf{y}$ can be written as

$$P(\mathbf{y}|\mathbf{x}, \mathbf{a}) = \begin{cases} a_n \cdot \lambda_n \cdot \tau(\mathbf{x}, \mathbf{a}), & \text{if } \mathbf{y} = \mathbf{x} + (1, 0, 0), \\ a_s \cdot \lambda_s \cdot \tau(\mathbf{x}, \mathbf{a}), & \text{if } \mathbf{y} = \mathbf{x} + (0, 1, 0), \\ a_h \cdot \lambda_h \cdot \tau(\mathbf{x}, \mathbf{a}), & \text{if } \mathbf{y} = \mathbf{x} + (0, 0, 1), \\ x_n \cdot \mu_n \cdot \tau(\mathbf{x}, \mathbf{a}), & \text{if } \mathbf{y} = \mathbf{x} - (1, 0, 0), \quad (8.45) \\ x_s \cdot \mu_s \cdot \tau(\mathbf{x}, \mathbf{a}), & \text{if } \mathbf{y} = \mathbf{x} - (0, 1, 0), \\ x_h \cdot \mu_h \cdot \tau(\mathbf{x}, \mathbf{a}), & \text{if } \mathbf{y} = \mathbf{x} - (0, 0, 1), \\ 0, & \text{if } \mathbf{y} = \mathbf{x}. \end{cases}$$

It is also shown in Fig. 8.1.

8.3.2 Solution via Linear Programming

By using the uniformization technique, we can derive a modified linear programming associated with the MDP for the maximum reward

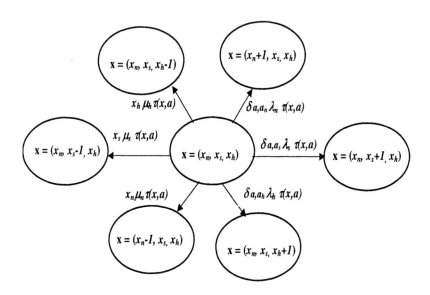

Figure 8.1: Illustration of the transition probability $P(\mathbf{y}|\mathbf{x}, \mathbf{a})$.

function. It is given below with decision variables $\pi(\mathbf{x}, \mathbf{a})$, $\mathbf{x} \in X$ and $\mathbf{a} \in A_{\mathbf{x}}$.

$$\text{Maximize} \sum_{\mathbf{x} \in X} \sum_{\mathbf{a} \in A_{\mathbf{x}}} \mathrm{r}(\mathbf{x}, \mathbf{a}) \tau(\mathbf{x}, \mathbf{a}) \pi(\mathbf{x}, \mathbf{a}) \tag{8.46}$$

subject to

$$\sum_{\mathbf{a} \in A_y} \pi(\mathbf{y}, \mathbf{a}) - \sum_{\mathbf{x} \in X} \sum_{\mathbf{a} \in A_x} P(\mathbf{y}|\mathbf{x}, \mathbf{a}) \pi(\mathbf{x}, \mathbf{a}) = 0, \forall \ \mathbf{y} \in X \tag{8.47}$$

$$\sum_{\mathbf{x} \in X} \sum_{\mathbf{a} \in A_x} \tau(\mathbf{x}, \mathbf{a}) \pi(\mathbf{x}, \mathbf{a}) = 1, \tag{8.48}$$

$$\pi(\mathbf{x}, \mathbf{a}) \geq 0, \quad \mathbf{x} \in X, \quad \mathbf{a} \in A_{\mathbf{x}}. \tag{8.49}$$

The variables $\pi(\mathbf{x}, \mathbf{a})$ satisfying Eqs. (8.47)-(8.49) can be viewed as the steady-state probabilities of being in state $\mathbf{x}$ and choosing action $\mathbf{a}$. The optimal action in each state can be chosen among all actions in each state since the value of $\pi(\mathbf{x}, \mathbf{a})$ will be zero for all but one action in each state. This implies that the optimal policy is non-randomized admission policy and the action $\mathbf{a}$ chosen for each state is a deterministic function of state $\mathbf{x}$.

For every given state, we use the linear programming technique to compute the optimal value of $\pi(\mathbf{x}, \mathbf{a})$ that maximizes the reward function as given in Eq. (8.48). Then, an optimal action for each state can be determined by comparing values of $\pi(\mathbf{x}, \mathbf{a})$ among actions of each state. These actions can be then tabulated a CAC policy, which is a mapping function from system states to actions. The policy is implemented in the OPNET simulator as a lookup table so that one can make the proper action based on the system state and the occurrence of the call arrival or departure event. Similar to 2D case, Fig. 8.9 also shows the implementation steps for the MDP-LP policy (Markov Decision Process and Linear Program), it also shows the performance comparison of the MDP-LP policy with various schemes, *i.e.* the complete sharing (CS) scheme and the GC schemes using OPNET simulator.

8.4   Simulation Results and Discussion

8.4.1   Numerical Results for 2-D CAC Policy using MDP

In this section, the CAC policy on two types call is investigated. Compared with previous work in [15, 17], our work is focused on how the optimal CAC policy varies with traffic parameters. The optimal CAC policy is illustrated (with the total system capacity equal to a bandwidthe of 10 units) in Figs. 8.2 to 8.5 for light and heavy traffic scenarios, respectively.

Light Traffic Scenario

If the total service rate is larger than the total arrival rate, the system is in the light traffic scenario. In such a case, the CAC policy rejects to accept calls only when the system exceeds its capacity limit. The resulting CAC policy performs similarly to the "complete sharing scheme" as shown in Figs. 8.2(a), (b) and Fig. 8.3(a). In Fig. 8.2(a), service rates $\mu_n = \mu_h = 1$ are greater than $\lambda_n = \lambda_h = 0.5$.

Similarly, in Fig. 8.2(b), the service rates $\mu_n = \mu_h = 2$ are greater than $\lambda_n = \lambda_h = 1$. In Fig. 8.3(a), the service rates $\mu_n = \mu_h = 1$ are greater than $\lambda_h = 0.5$.

Because of the use of weighting factors $w_n = 1$ and $w_h = 10$, the dominant parameter is the handoff arrival rate λ_h. The comparison between Fig. 8.3(a) and (b) provides a clear proof. In these two cases, their service rates are kept the same as those in Fig. 8.2(a), *i.e.* $\mu_n = \mu_h = 1$. Even though the new call arrival rate λ_n is increased to 2 in Fig. 8.3(a), a CAC policy similar to that in (a) is obtained due to the unchanged handoff arrival rate ($\lambda_h = 0.5$), which is the dominant arrival parameter. On the other hand, if λ_h, the dominant parameter, is increased as shown in Fig. 8.3(b), the optimal CAC policy becomes the guard channel type as discussed later.

Heavy Traffic Scenario

If the total service rate is smaller than the total arrival rate, the system is in the heavy traffic scenario, and the optimal CAC policy should be designed based on the GC scheme as shown in Figs. 8.4(a) and (b). In Fig. 8.4(a), service rates $\mu_n = \mu_h = 0.5$ are less than arrival rates $\lambda_n = \lambda_h = 1$. Similarly, in Fig. 8.2(b), service rates $\mu_n = \mu_h = 1$ are less than arrival rates $\lambda_n = \lambda_h = 3$.

It is worthwhile to point out that, because of the use of weighting factors $w_n = 1$ and $w_h = 10$, the dominant parameter is the handoff arrival rate λ_h. Fig. 8.5(a) and (b) confirms this claim, where the service rates $\mu_n = \mu = 1$ and the dominant parameter $\lambda_h = 3$ are kept the same as those in Fig. 8.4(b). The increase of the new call arrival rate, as done in Fig. 8.5(a) with $\lambda_n = 5$ or the decrease of it, as done in Fig. 8.5(b) with $\lambda_n = 0.8$, does not have an impact on the guard channel nature of the CAC policy. The advantage of using MDP is that MDP provides an accurate number of guard channels and a finer level of policy adjustment.

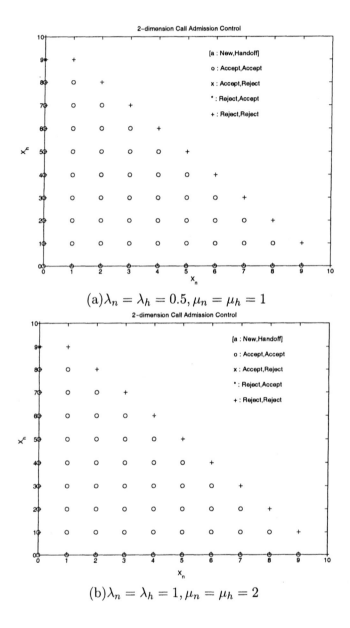

(a)$\lambda_n = \lambda_h = 0.5, \mu_n = \mu_h = 1$

(b)$\lambda_n = \lambda_h = 1, \mu_n = \mu_h = 2$

Figure 8.2: Optimal CAC policies in the light traffic scenario for (a)$\lambda_n = \lambda_h = 0.5, \mu_n = \mu_h = 1$ and (b)$\lambda_n = \lambda_h = 1, \mu_n = \mu_h = 2$.

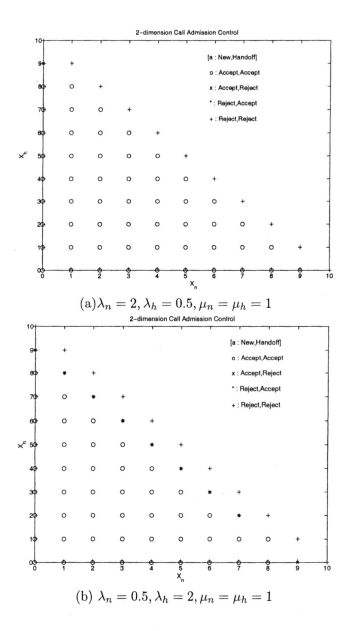

(a)$\lambda_n = 2, \lambda_h = 0.5, \mu_n = \mu_h = 1$

(b) $\lambda_n = 0.5, \lambda_h = 2, \mu_n = \mu_h = 1$

Figure 8.3: Optimal CAC policies in the light traffic scenario for (a)$\lambda_n = 2, \lambda_h = 0.5, \mu_n = \mu_h = 1$ and (b) $\lambda_n = 0.5, \lambda_h = 2, \mu_n = \mu_h = 1$.

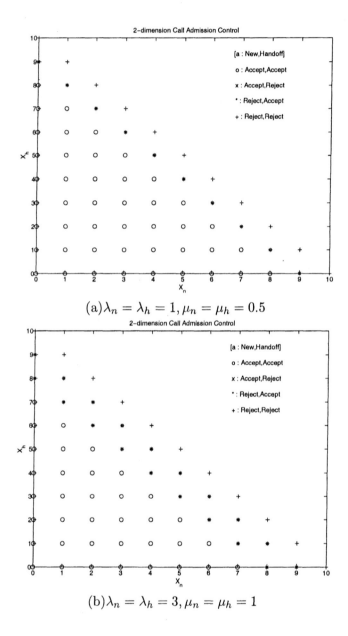

Figure 8.4: Optimal CAC policies in the heavy traffic scenario for (a)$\lambda_n = \lambda_h = 1, \mu_n = \mu_h = 0.5$ and (b)$\lambda_n = \lambda_h = 3, \mu_n = \mu_h = 1$.

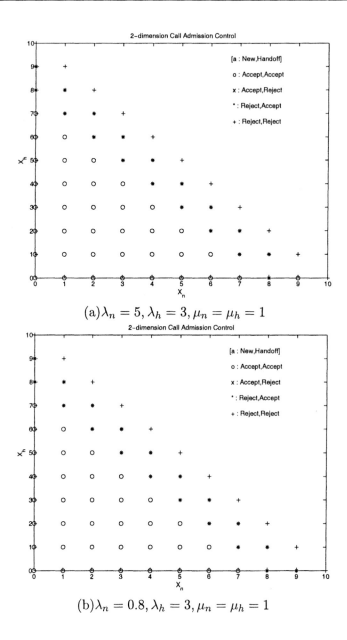

Figure 8.5: Optimal CAC policies in the heavy traffic scenario for (a)$\lambda_n = 5, \lambda_h = 3, \mu_n = \mu_h = 1$ and (b)$\lambda_n = 0.8, \lambda_h = 3, \mu_n = \mu_h = 1$.

Discussion on Relationship between CAC Policies and Traffic Conditions

Figs. 8.6 and 8.7 show how the optimal CAC policy changes with a different handoff service rate μ_h under the heavy traffic condition.

Sufficient Resource As shown in Fig. 8.6(a), faster service rates (larger μ_h and μ_n) help the system to have enough resource even in a heavy traffic situation. A larger number of μ_h increases the system reward more efficiently due to a higher weight of the handoff service rate μ_h on the reward function. As a result, the system blocks both calls only when all resources are in use.

In-sufficient Resource Figs. 8.6(b) to 8.7 (a) then to (b) show the evolution of slowing down the handoff service rate μ_h from 3 to 2 and then to 1. Smaller service rates result in longer system occupation time and reduce the average residual system resource. The effect of slower handoff service rate μ contributes to a lower reward function. As the service rate decreases, the CAC policy will change according to the following two rules to optimize the system reward under the heavy traffic situation.

- *Rule* 1 (General Trend of the CAC Policy):

    When the resource becomes in-sufficient due to heavy arrivals and/or slow service rates, the general trend of the CAC policy is to have two types of states: (1) reject all calls when system states satisfy $\mathbf{x}=\{(x_n, x_h)|x_n + x_h = C = 10\}$, and (2) accept only handoff calls for states $\mathbf{x}=\{(x_n, x_h)|x_n + x_h = i, t \leq i \leq 9\}$, where parameter t is called the CAC policy trend parameter.

- *Rule* 2 (Fine-Tuning of the CAC Policy):

    On top of the general trend, we also observe some fine tuning in the CAC policy according to the number of pre-occupied handoff calls. To give an example, let us focus on those states satisfying $\mathbf{x}=\{(x_n, x_h)|x_n + x_h = t = 9\}$ as shown in Fig. 8.6(b) ($\mu_h = 3$). States $(x_n, x_h) = (4, 5)$, $(3, 6)$ have a different CAC policy. This state-dependent

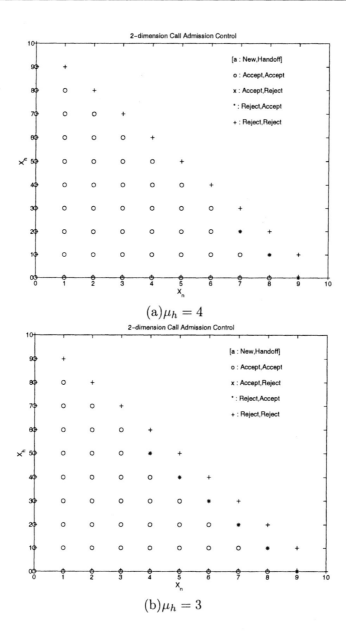

Figure 8.6: Optimal CAC policies for (a) $\mu_h = 4$, (b) $\mu_h = 3$ with the total system capacity equal to $C = 10$ unit bandwidth, $\lambda_n = \lambda_h = 4$ and $\mu_n = 1$.

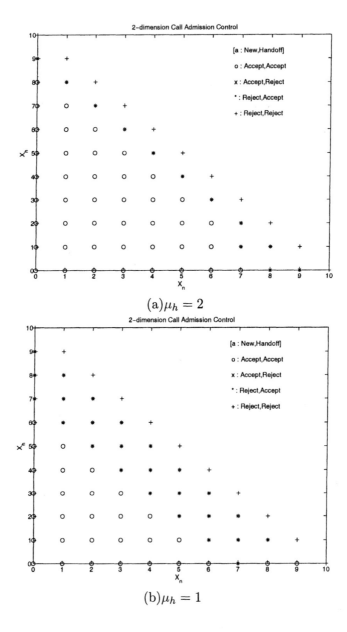

Figure 8.7: Optimal CAC policies for (a) $\mu_h = 2$, (b) $\mu_h = 1$, with the total system capacity equal to $C = 10$ unit bandwidth, $\lambda_n = \lambda_h = 4$ and $\mu_n = 1$.

CAC policy only occurs with the MDP model and will not be allowed in traditional GC schemes.

If all parameters are fixed except new call service rate μ_n, the CAC policy follows the same general trend, but differs in the fine-tuning rule along boundary states. For those states satisfying $\mathbf{x} = \{(x_n, x_h) | x_n + x_h = t = 8\}$ as shown in Fig. 8.8(a), states $(x_n, x_h) = (6, 2)$ and $(5, 3)$ have a different policy since the CAC policy would like to reject new calls only when the system is not served fast enough (no enough resource). With the same total number of occupied calls, the former state $(6, 2)$ consists of more fast service new calls and, therefore, the system resource in this state is considered sufficient to accept all calls.

For every given state, we use the linear programming technique to compute the optimal value of $\pi(\mathbf{x}, \mathbf{a})$ that maximizes the reward function as given in (8.46). Then, an optimal action for each state can be determined by comparing values of $\pi(\mathbf{x}, \mathbf{a})$ among actions of each state. These actions can be then tabulated a CAC policy, which is a mapping function from system states to actions. The policy is implemented in the OPNET simulator as a lookup table so that one can make the proper action based on the system state and the occurrence of the call arrival or departure event. Fig. 8.9 shows the implementation steps for the MDP-LP policy (Markov Decision Process and Linear Program). It also shows the performance comparison of the MDP-LP policy with other two schemes, *i.e.* the complete sharing (CS) scheme and the GC scheme with OPNET.

Performance Comparison

Let's describe the traffic parameters used in simulation model first. The Poisson call arrival rate and the exponentially distributed call holding time are assumed in the experiment. The call arrival rate is controlled directly by λ_n and λ_h for the new call and hard handoff arrivals, respectively. The call holding time for a new call and a hard handoff are directly controlled by $1/\mu_n$ and μ_h, respectively. A large value of λ results in the increment of the network traffic load. Values used in the simulation models are listed below.

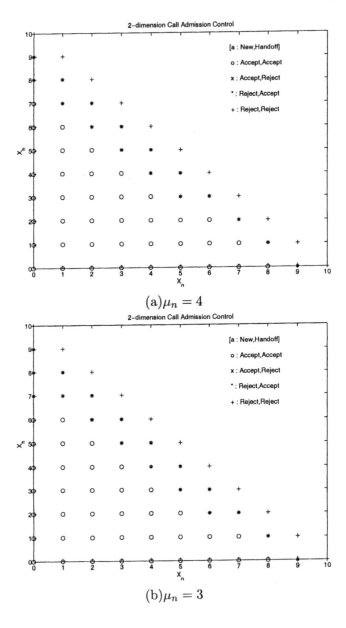

$$(a) \mu_n = 4$$

$$(b) \mu_n = 3$$

Figure 8.8: Optimal CAC policy under heavy traffic with different new call service rates μ_n: (a) $\mu_n = 4$ and (b) $\mu_n = 3$, where the total system capacity $C = 10$, $\lambda_n = \lambda_h = 4$ and $\mu_h = 1$ are fixed.

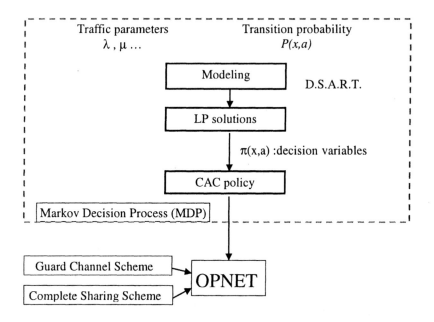

Figure 8.9: Implementation steps for the MDP-LP CAC policy.

- $\lambda_n = 4$: the arrival rate of new calls;
- $\lambda_h = 3.1$: the arrival rate of hard handoff calls;
- $\mu_n = 1$: the service rate of new calls;
- $\mu_h = 1.4$: the service rate of hard handoff calls;
- $C = 10$: the total system bandwidth units;
- $w_n = 1$, $w_h = 10$: weighting factors for new call and hard handoff.

We compare the performance of MDP-LP with various guard channel schemes $GC(TH)$ with different threshold value $TH = i$, $i \in 0, 1, 2, 3$. Let us describe the call admission control policy in $GC(TH)$ scheme as follows. If there is less or equal to TH channels left, only hard handoff calls can be admitted. The complete sharing scheme is equivalent to the guard channel scheme with 0 unit guard channel, which is denoted as GC(0). For the CDMA system, the "channel" is replaced by the term "unit bandwidth".

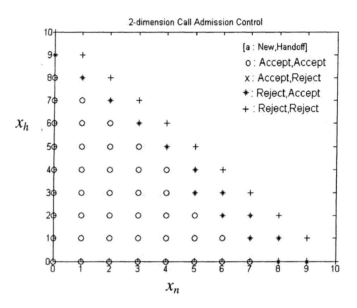

Figure 8.10: The MDP CAC policy for the 2D Case.

To approximate the definition given by (8.50), the QoS metric is measured by the average weighted reward ϕ which is defined as

$$\phi \;=\; \frac{1}{N_e} \sum_{n=1}^{N_e} r(\mathbf{x}, \mathbf{a})$$

$$\;=\; \frac{1}{N_e} \sum_{n=1}^{N_e} \{w_n(a_n \cdot e_n + x_n) + w_h(a_h \cdot e_h + x_h)\}, \quad (8.50)$$

where N_e is the total number of events in the simulation, including all arrival events $\mathbf{e} = (e_n, e_h)$, $r(\mathbf{x}, \mathbf{a})$ is the reward function for the current event, $\mathbf{x} = (x_n, x_h)$ and $\mathbf{a} = (a_n, a_h)$ are the current state and the action chosen by CAC, respectively, where $0 \leq x_n, x_h \leq C$ and a_n and a_h take the binary value.

The optimal policy derived by the MDP-LP decision process is shown in Fig. 8.10. Each point represents one possible system state of $\mathbf{x} = (x_n, x_h)$, $0 \leq x_n, x_h \leq C$. The action vector $\mathbf{a} = (a_n, a_h)$ chosen for each state is represented by a different symbol. In this example, four possible actions are found in all coordinates. If action $\mathbf{a} = (0,0)$ is chosen (symbol '+'), all types of calls are rejected. This happens at coordinates of $\mathbf{x} = \{(x_n, x_h)|x_n + x_h = C\}$. If action $\mathbf{a} = (0,1)$ is chosen (symbol '*'), only the hard handoff call is accepted. Similarly, action $\mathbf{a} = (1,1)$ (symbol 'o') denotes accepting all types of calls, and action $\mathbf{a} = (1,0)$ (symbol 'x') denotes accepting the new call only.

The average weighted reward, as defined in (8.52), for different schemes are shown in Figs. 8.11(a) and (b). We see that the MDP-LP decision rule gives the best result when compared with all guard channels with various thresholds. Fig. 8.11(b) gives a close look at the performance comparison among MDP-LP, GC(1) and GC(2) schemes.

The new call blocking and the handoff dropping probabilities under different control schemes are shown in Figs. 8.12(a) and (b), respectively. Results show that the proposed MDP-LP scheme optimizes the reward function using the optimal CAC policy for each system state. Such an optimal CAC policy provides a better controlling resolution that cannot be achieved by the GC scheme only.

8.4.2   Numerical Results for 3-D CAC Policy using MDP

Traffic Parameters

The Poisson call arrival rate and the exponentially distributed call holding time are assumed in the experiment. The call arrival rate is controlled directly by λ_n, λ_h and λ_s for the new call, hard handoff and soft handoff arrivals, respectively. The λ values are in the units of calls per minute. The mean request arrival rate is measured in the number of connections per minute. The call holding time for a new call is directly controlled by $1/\mu_n$. A large value of λ results in the increment of the network traffic load. The call holding time for hard and soft handoff calls is controlled by μ_h and μ_s, which are the service rate for hard and soft handoff calls, respectively.

Values used in the simulation models are listed below.

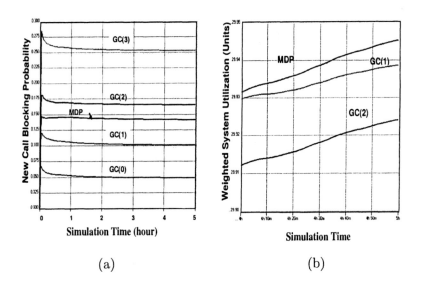

Figure 8.11: The performance comparison by weighted reward (a) MDP-LP, GC(0), GC(1), GC(2) and GC(3) schemes, (b) a close look at the performance among MDP-LP, GC(1) and GC(2) schemes.

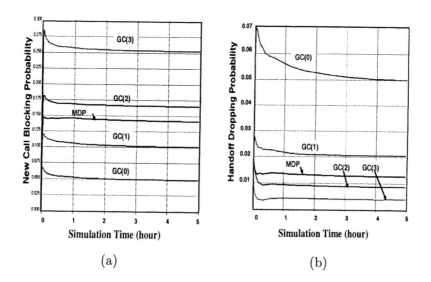

Figure 8.12: (a) The new call blocking probabilities for various schemes, and (b) handoff dropping probabilities for various schemes.

- $\lambda_n = 2.5$: the arrival rate of new calls;
- $\lambda_s = 2.5$: the arrival rate of soft handoff calls;
- $\lambda_h = 2.5$: the arrival rate of hard handoff calls;
- $\mu_n = 0.5$: the service rate of new calls;
- $\mu_s = 1$: the service rate of soft handoff calls;
- $\mu_h = 1.5$: the service rate of hard handoff calls;
- $C = 10$: the total system bandwidth units;
- $w_n = 1$, $w_s = 5$, $w_h = 10$: weighting factors for new, soft handoff and hard handoff calls.

OPNET Implementation

Simulations were conducted by using the OPNET simulator [22]. The CAC policy was evaluated in a distributed manner.

Fig. 8.13 shows the simulation system. Three call generating processes are deployed to generate traffics for new calls, soft handoff calls and hard handoff calls, respectively. The resource management module is responsible for resource management, where call admission control (CAC) is included in this module. Fig. 8.14(a) and (b) illustrate the call generator process and the resource management process, respectively. Detail state transition diagrams for them are depicted in these figures.

For each call type, a call generation process is generating calls according to Poission distribution with a given traffic parameters as shown in Fig. 8.14(a). Each call's birth and death is controlled by the call generating process while the call admission control is governed by the resource management process as shown in Fig. 8.14(b) according to the CAC policy such as the complete sharing scheme, the guard channel scheme and the optimal control policy.

Events and states are listed below to describe the finite state machine of the resource management process. In addition to the IDLE state, there are six states: (1) the NEW_CAC state, (2) the CAC_SOFT_HANDOFF state, (3) the CAC_HARD_HANDOFF state, (4) the NEW_CALL_TERM state, (5) the DEPART_SOFT_HANDOFF state, and (6) the DEPART_HARD_HANDOFF state. Events trigger the resource management process to transit from one state to the other as explained below.

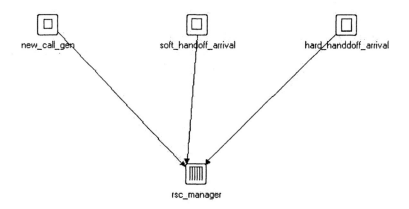

Figure 8.13: The OPNET simulation system.

- Event NEW_ARRIVAL occurs when a new call is requesting access to the target cell. This event will trigger the transition from the IDLE state to the NEW_CAC state. The admission control for the new call will then be performed. The system returns to the IDLE state after the admission decision is made.

- Event SOFT_ARRIVAL occurs when a soft handoff call is requesting access to the target cell. This event will trigger the transition from the IDLE state to the CAC_SOFT_HANDOFF state. The admission control for soft handoff call will then be performed. The system returns to the IDLE state after the admission decision is made.

- Event HARD_ARRIVAL occurs when a hard handoff call is requesting access to the target cell. This event will trigger the transition from the IDLE state to the CAC_HARD_HANDOFF state. The admission control for hard handoff call will then be performed. The system returns to the IDLE state after the admission decision is made.

- Event NEW_TERM occurs when a new call is terminating.

This event will trigger the transition from the IDLE state to the NEW_CALL_TERM state. The system resource used by that new call will be released. The system returns to the IDLE state afterwards.

- Event SOFT_DEPART occurs when a soft handoff call is handed off to another cell or terminates. This event will trigger the transition from the IDLE state to the DEPART_SOFT_HANDOFF state. The system resource used by that soft handoff call will be released. The system returns to the IDLE state afterwards.

- Event HARD_DEPART occurs when a hard handoff call is handed off to another cell or terminates. This event will trigger the transition from the IDLE state to the DEPART_HARD_HANDOFF state. The system resource used by that hard handoff call will be released. The system returns to the IDLE state afterwards.

Performance Comparison

Let us compare the performance of MDP-LP with two other schemes, *i.e.* the complete sharing scheme (CS) and the guard channel scheme with two thresholds, denoted by $GC(TH_1, TH_2)$. In the simulation, we used $TH_1 = 1$ and $TH_2 = 3$ or $GC(1,3)$ for comparison. With a total of $C = 10$ unit bandwidths, the call admission scheme of $GC(1,3)$ does not block any types of calls until there are 3 channels left. If there are less or equal to $TH_2 = 3$ channels left, only soft handoff and hard handoff calls can be admitted. If there is less or equal to $TH_1 = 1$ channel left, only hard handoff calls can be admitted. For the CDMA system, the "channel" is replaced by the term "unit bandwidth".

To approximate the definition given by (8.11), the QoS metric is measured by the average weighted reward ϕ which is defined as

$$
\begin{aligned}
\phi &= \tfrac{1}{N_e} \textstyle\sum_{n=1}^{N_e} r(\mathbf{x}, \mathbf{a}) \\
&= \tfrac{1}{N_e} \textstyle\sum_{n=1}^{N_e} \{w_n(a_n \cdot e_n + x_n) + w_s(a_s \cdot e_s + x_s) + w_h(a_h \cdot e_h + x_h
\end{aligned}
$$
(8.51)

where N_e is the total number of events in the simulation, including all arrival events $\mathbf{e} = (e_n, e_s, e_h)$, $r(\mathbf{x}, \mathbf{a})$ is the reward function

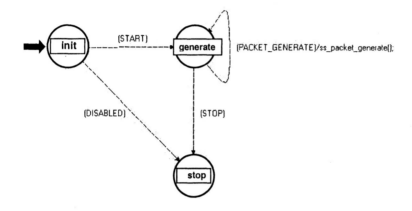

(a) The call generating process

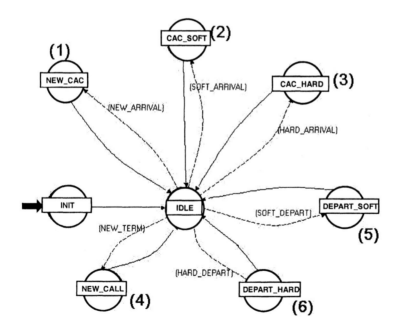

(b) The resource management process

Figure 8.14: The OPNET simulated processes.

for the current event, $\mathbf{x} = (x_n, x_s, x_h)$ and $\mathbf{a} = (a_n, a_s, a_h)$ are the current state and the action chosen by CAC, respectively, where $0 \leq x_n, x_s, x_h \leq C$ and a_n, a_s and a_h take the binary value.

The optimal policy derived by the MDP-LP decision process is shown in Fig. 8.15. Each point represents one possible system state of $\mathbf{x} = (x_n, x_s, x_h)$, $0 \leq x_n, x_s, x_h \leq C$. The action vector $\mathbf{a} = (a_n, a_s, a_h)$ chosen for each state is represented by a different symbol. In this example, four possible actions are found in all coordinates. If action $\mathbf{a} = (0, 0, 0)$ is chosen (symbol 'x'), all types of calls are rejected. This happens at coordinates of $\mathbf{x} = \{(x_n, x_s, x_h)|x_n + x_s + x_h = C\}$. If action $\mathbf{a} = (0, 1, 1)$ is chosen (symbol '*'), only soft and hard handoff calls are accepted. Similarly, action $\mathbf{a} = (1, 1, 1)$ (symbol 'o') denotes accepting all types of calls, and action $\mathbf{a} = (0, 0, 1)$ (symbol '+') denotes accepting hard handoff only.

The normalized average weighted reward, as defined in Eq. (8.50), is shown in Fig. 8.16. We see that the MDP-LP decision rule gives better results than the complete sharing (CS) scheme and the GC(1,3) guard channel scheme. As mentioned in Section 8.1, we can view the multiple-threshold guard channel scheme as a control subset of the proposed high resolution decision rule based on system states, whose solution was obtained by linear programming and proved to be optimal as done in [12] and [23].

The dropping probability for each class under different control schemes is shown in Fig. 8.17. Fig. 8.17(a) shows that the blocking probabilities are at a similar level because system resources are completely shared among all types of users. Fig. 8.17(b) shows that the GC(1,3) guard channel scheme provides a preferential treatment to hard and soft handoff users. Such a preferential treatment reserves more resources for the handoff usage so that it leaves a part of the system idle in some cases if not controlled carefully. On the other hand, the proposed scheme as shown in Fig. 8.17(c) optimizes the reward function well using the optimal CAC policy for each system state.

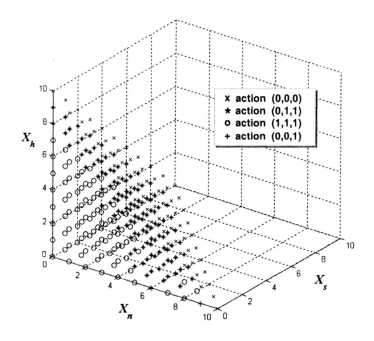

Figure 8.15: The MDP CAC policy.

8.5 Complexity of the MDP using Linear Programming Approach

The complexity of the Markov Decision Process was discussed in literature [24], [25], [26], [27],[28] and [29]. In this section, we focus on the complexity of the Linear Programming method. We consider two parameters related to the computation complexity: (1) the number of total states $N_{\mathbf{x}}$, and (2) the number of total decision variables N_{π}. Since the optimal CAC policy for each state is obtained from the Linear Programming solution by referencing the value of decision variables N_{π} determines the problem size and the computation complexity of the MDP problem.

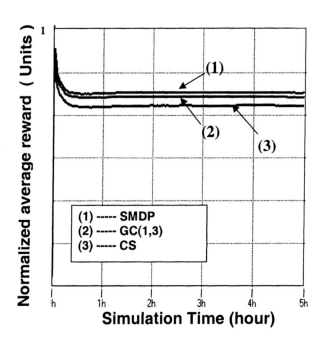

Figure 8.16: The average reward function.

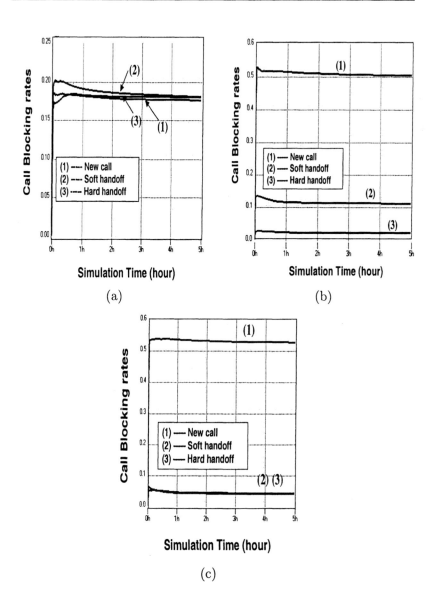

Figure 8.17: The blocking probabilities for (a) the complete sharing scheme, (b) the GC(1,3) guard channel scheme and (c) the MDP stationary policy.

8.5.1   Complexity Issue for Linear Programming

Let X denote the state space consisting of k types of calls in the system, *i.e.*

$$X = \left\{ \mathbf{x} | \mathbf{x} = (x_1, x_2, \cdots, x_k); x_1, \cdots, x_k \geq 0, \sum_{i=1}^{k} x_i \leq C \right\}, \quad (8.52)$$

where x_i is the number of calls for call type i, and the maximum system capacity can be expressed as $C = \lfloor \frac{(W/R)(1-\eta)}{\epsilon} \rfloor$. The action space $A_{\mathbf{x}}$ for state $\mathbf{x} \in X$ can be written as

$$A_{\mathbf{x}} = \begin{cases} \{\mathbf{a} = (1, 1, \cdots, 1)\}, & \text{if } \mathbf{x} = (0, 0, \cdots, 0), \\ \{\mathbf{a} = (0, 0, \cdots, 0)\}, & \text{if } \sum_{i=1}^{k} x_i = C, \\ \{\mathbf{a} | \mathbf{a} = (a_1, a_2, \cdots, a_k); & \text{Otherwise} \\ \quad a_1, a_2, \cdots, a_k \in \{0, 1\}\}, \end{cases} \quad (8.53)$$

Let $N_{\mathbf{a}}(\mathbf{x})$ denote the total number of actions for state $\mathbf{x}$. It can be written as

$$N_{\mathbf{a}}(\mathbf{x}) = \begin{cases} 1, & \text{if } \mathbf{x} = (0, 0, \cdots, 0), \\ 1, & \text{if } \sum_{i=1}^{k} x_i = C, \\ 2^k, & \textbf{Otherwise}. \end{cases} \quad (8.54)$$

Following the same notation, let us first define the following vectors. The size of all following vectors is $N_\pi \times 1$. For the purpose of indexing, let r_i, τ_i, c_i and π_i be the i-th indexed elements in $\underline{r}$, $\underline{\tau}$, $\underline{c}$ and $\underline{\pi}$, respectively. Each is associated with possible $(\mathbf{x}, \mathbf{a})$ values. Thus, we have

$$\begin{aligned} \underline{r} &= \left[r_1, r_2, \cdots, r_i, \cdots, r_{N_\pi} \right]^T \\ \underline{\tau} &= \left[\tau_1, \tau_2, \cdots, \tau_i, \cdots, \tau_{N_\pi} \right]^T \\ \underline{c} &= -\left[r_1 \tau_1, r_2 \tau_2, \cdots, r_i \tau_i, \cdots, r_{N_\pi} \tau_{N_\pi} \right]^T \\ \underline{\pi} &= \left[\pi_1, \pi_2, \cdots, \pi_i, \cdots, \pi_{N_\pi} \right]^T \end{aligned}$$

We can map our problem in (8.19)- (8.22) into to a standard-form linear program as

$$\begin{aligned} \textbf{LP}: \quad &min \quad \underline{c}^T \underline{\pi} \\ &\text{s.t.} \quad \mathbf{G}\underline{\pi} = \underline{b}, \\ &\qquad \underline{\pi} \geq \underline{0}. \end{aligned}$$

where $\underline{b}_{(N_x+1)\times 1} = [0,0,\cdots,0,1]^T$, and matrix $\mathbf{G}_{(N_x+1)\times N_\pi}$ can be defined as the following form:

$$\mathbf{G}_{(N_x+1)\times N_\pi} = \left[\frac{\mathbf{P}_{N_x\times N_\pi} - \mathbf{D}_{N_x\times N_\pi}}{\underline{c}^T_{1\times N_\pi}} \right] \tag{8.55}$$

where $\mathbf{P}_{N_x\times N_\pi}$ is the state transition probability matrix with elements of $P(\mathbf{y}|\mathbf{x},\mathbf{a})$ defined in Eq. (8.13). $\mathbf{D}_{N_x\times N_\pi}$ is an block diagonal matrix with diagonal blocks l_{ii} of all ones, and zeros for non-diagonal blocks $l_{ij}, \forall i \neq j$. Thus,

$$\mathbf{D}_{N_x\times N_\pi} = \begin{bmatrix} l_{11} & \underline{0} & \cdots & \cdots & \cdots & \cdots \\ \underline{0} & l_{22} & \underline{0} & \cdots & \cdots & \cdots \\ \cdots & \cdots & \ddots & \cdots & \cdots & \cdots \\ \cdots & \cdots & \underline{0} & l_{ii} & \underline{0} & \cdots \\ \cdots & \cdots & \cdots & \cdots & \ddots & \cdots \\ \cdots & \cdots & \cdots & \underline{0} & l_{N_xN_x} \end{bmatrix} \tag{8.56}$$

where if there are k types of calls, the diagonal block l_{ii} is a vector of all ones of the following dimension,

$$l_{ii} = \begin{cases} \underline{1}_{1\times 1}, & \text{if indexed i-th state satisfies} \\ & \mathbf{x} = (0,0,\cdots,0), \text{or } \sum_{n=1}^k x_n = C, \\ \underline{1}_{1\times 2^k}, & \text{Otherwise.} \end{cases} \tag{8.57}$$

A well known solution using the Interior Point Method was proposed by Karmarkar [26]. This algorithm solves an LP problem in the polynomial time, and has been adopted by Matlab for large-scale LP problems. The complexity merits for this algorithm are usually expressed as an upper bound of the magnitude of the number of arithmetic operations, and they are a function of the problem size. One can interpret the *size* as a measure of the computation loading involved.

It is obvious that computation loadings are different if the elements in $\mathbf{G}$, $\underline{\pi}$, and $\underline{b}$ are different. The more 0 and 1 elements in equations, the less computation involved. In order to capture the problem size denoted by L, it requires an agreed upon definition, which results in rather subtle work than just the number of decision variables (N_π) and constraints $(N_x + 1)$. One commonly accepted measure of the size of an LP problem is $L \simeq (N_x+1) \cdot N_\pi + \lceil \log_2 |v| \rceil$, where v is the product of all of nonzero elements in the problem (in matrix $\mathbf{G}$, vector $\underline{\pi}$, and $\underline{b}$) [20]. For this measure, these matrices and vectors are assumed to contain only integers (which can always be achieved by appropriate scaling) and $\lceil \log_2 |v| \rceil$ reflects the number of bits needed to represent the problem in binary notation for a computer.

A numerical illustration of the problem size can be found on pages 494-496 in Miller's book [20]. Karmarkar's algorithm can solve the LP problem in the order of $O(N_\pi^4 L)$, where L is the size of the problem. The overall complexity for Karmarkar's basic algorithm requires $O(N_\pi^4 L)$ operations in "big-O" notation. Here, the operation is referred to as arithmetic operations and comparisons in infinite precision as described by Anstreicher [30], in which a comprehensive survey on the complexity issue for other algorithms was presented. We summarized his comparison results in Table 8.1.

Table 8.1: Complexity for Solving LP

Algorithm	Complexity	Authors
Projective (basic)	$O(N_\pi^4 L)$	Karmarkar [26]
Projective partial updating	$O(N_\pi^{3.5} L)$	Karmarkar [26]
Path following	$O(N_\pi^{3.5} L)$	Renegar [27]
Partial updating + Path following	$O(N_\pi^3 L)$	Gonzaga [28] and Vaidya [29]
Partial updating + Preconditioned conjugate gradient	$O(\frac{N_\pi^3}{\ln N_\pi} L)$	Anstreicher [30]

8.5.2   Derivation of the Problem Size

The 2D Case

In a system with two call types and of capacity C, and the total number of actions for each state can be represented as in Eq. (8.54), in which $k = 2$. The total number of decision variables $N_{\mathbf{x}}$ can be derived as

$$N_{\mathbf{x}} = 1 + 2 + \cdots + (C + 1) = \frac{(C + 1)(C + 2)}{2}. \tag{8.58}$$

Fig. 8.18 shows the state diagram of the 2-D MDP model, where each node stands for a state. Next, we determine the number of decision variables in the model. To reduce the number of decision variables, we make two reasonable assumptions for boundary states:

- At state $\mathbf{x} = (0,0)$, accept all call types $i.e.$, $\mathbf{a} = (1,1)$.

- At states with $x_n + x_h = C$, reject all incoming calls $i.e.$ $\mathbf{a} = (0,0)$.

Thus, only one decision variable is allowed for each boundary state. For the remaining states, there are $2^2 = 4$ decision variables, each for one possible action in that state. The total number of decision variables can be computed by

$$N_{\pi} = 1 + \left(\frac{(C)(C + 1)}{2} - 1 \right) \cdot 2^2 + (C + 1). \tag{8.59}$$

Table 8.2 shows the values of $N_{\mathbf{x}}$, N_{π} with different system capacities C.

The 3D Case

In a system with two call types and of capacity C, and the total number of actions for each state can be represented as in Eq. (8.54), in which $k = 3$. The total number of decision variables $N_{\mathbf{x}}$ can be derived as

$$
\begin{aligned}
N_{\mathbf{x}} &= \text{Number of states on surface } S(x_h = C) \\
&+ \text{Number of states on surface } S(x_h = C - 1)
\end{aligned}
$$

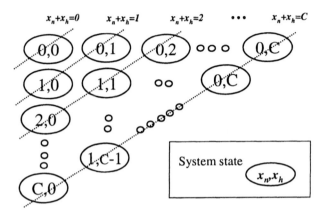

Figure 8.18: The computation of the total number of states N_x in a 2-D MDP model.

Table 8.2: Total number states and decision variables in a 2-D MDP model

C	N_x	N_π	C	N_x	N_π
5	21	63	30	496	1888
10	66	228	40	861	3318
15	136	493	50	1326	5148
20	231	858	60	1891	7378

$$+ \quad \cdots\cdots$$

$$+ \quad \text{Number of states on surface } S(x_h = i)$$

$$+ \quad \cdots\cdots$$

$$+ \quad \text{Number of states on surface } S(x_h = 0)$$

$$= \frac{1 \cdot 2}{2} + \frac{2 \cdot 3}{2} + \cdots + \frac{(C+1) \cdot (C+2)}{2}$$

$$= \sum_{i=C}^{0} \frac{(C-i+1) \cdot (C-i+2)}{2}$$

$$= \frac{(C+1) \cdot (C+2) \cdot (C+3)}{6}, \tag{8.60}$$

Note that there are $(C - i + 1)(C - i + 2)/2$ states on the surface $S(x_h = i)$, $0 \le i \le C$. Fig. 8.19 gives the state diagram of a 3D MDP model, where each node stands for a state. Next, we determine the number of decision variables in the proposed model. In order to reduce the number of decision variables, we make two assumptions for boundary states:

- At state $\mathbf{x} = (0, 0, 0)$, accept all types of call, *i.e.* $\mathbf{a} = (1, 1, 1)$.

- At states where $x_n + x_s + x_h = C$, reject all incoming calls, *i.e.* $\mathbf{a} = (0, 0, 0)$.

Therefore, there is only one decision variable for each boundary state. For the remaining states, there are $2^3 = 8$ decision variables, each for one possible action in that state. As a result, the total number of decision variables is

$$N_\pi = 1 + \left(\frac{(C)(C+1)(C+2)}{6} - 1 \right) \cdot 2^3 + \frac{(C+1)(C+2)}{2}.$$

Table 8.3 shows the values of $N_{\mathbf{x}}$ and N_π under different system capacities.

Complexity Discussion for Higher Dimension

In a state with k call types and of capacity C, the total number of actions for each state is $N_{\mathbf{a}} = 2^k$ and the total number $N_{\mathbf{x}}$ of states is of $O(C^k)$. For real time implementation, a large number of C result

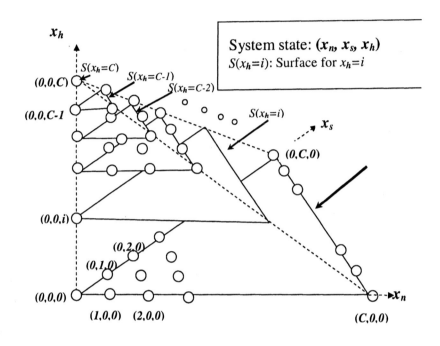

Figure 8.19: The total number of states N_x in a 3-D MDP model

Table 8.3: Total number states and decision variables in a 3-D MDP model

C	N_x	N_π	C	N_x	N_π
5	56	294	30	5456	40169
10	286	1819	40	12341	92694
15	816	5569	50	23426	178119
20	1771	12544	60	39711	304444

in a larger amount of memory and computational time requirements in a polynomial of order k. In the case of a higher dimension (say $k \geq 4$), an approximation model has to be developed to reduce the computational complexity. A good approximation model should find a good balance between accuracy and efficiency.

8.5.3   Conclusion and Future Work

Effective radio resource management schemes depend greatly on the CAC policy. The MDP was used to construct a system model, and the linear programming (LP) technique was used to find the optimal CAC policy at a given state in this chapter. We considered the CAC policies for a system with the hard handoff only or with a hybrid handoff process, which includes new, soft handoff, and hard handoff calls. The state concept in the traditional multiple threshold guard channel schemes can be viewed as a special case of the system state model discussed in this work. Our proposed scheme has a higher controlling precision. It was shown by computer simulation that the proposed scheme outperformed the complete sharing and the multiple threshold guard channel schemes.

Our work focused on the unconstrained Markov decision problem. The goal was to maximize the average weighted system reward without a constraint imposed on the dropping probability. If the dropping probability has a hard constraint, the CAC policy can be randomized as described in [11]. In other words, the policy in a specific state will choose a certain action based on a probability distribution rather than having a predicted action. In the future, it is worthwhile to consider CAC policies under some dropping probability constraints. The tradeoff between hard and soft handoff calls should also be studied.

BIBLIOGRAPHY

[1] K. Buchanan, R. Fudge, D. McFarlane, T. Phillips, A. Sasaki, and H. Xia, "IMT-2000: service provider's perspective," *IEEE Personal Commun.*, vol. 4 (4), pp. 8–13, 1997.

[2] R. Pandya, D. Grillo, E. Lycksell, P. Mieybegue, H. Okinaka, and M. Yabusaki, "IMT-2000 Standards: network aspects," *IEEE Personal Commun.*, vol. 4(4), pp. 20–29, 1997.

[3] D. Hong and S. S. Rapport, "Traffic model and performance analysis for cellular mobile radiotelephone systems with prioritized and nonprioritized handoff procedures," *IEEE Trans. on Vehicle Technology*, vol. 35, pp. 77–92, Aug. 1986.

[4] H. Chen, S. Kumar, and C.-C. J. Kuo, "Differentiated QoS aware priority handoff in cell-based multimedia wireless network," in *IS&T/SPIE's 12th Int. Symposium, Electronic Imaging 2000*, vol. 3974, (San Jose, CA), pp. 940–948, Jan. 2000.

[5] H. Chen, S. Kumar, and C.-C. J. Kuo, "Dynamic call admission control scheme for QoS priority handoff in multimedia cellular systems," in *IEEE Wireless Communications and Networking Conference*, vol. 1, (Orlando, Florida), pp. 114–118, March 2002.

[6] R. Ramjee, R. Nagarajan, and D. Towsley, "On optimal call admission control in cellular networks," in *IEEE INFOCOM'96*, pp. 43–50, 1996.

[7] C. W. Clark, "The lazy adaptable lions: a Markovian model of group foraging," *Anim. Behav.*, vol. 35, pp. 361–368, 1987.

[8] I. Houston and J. M. McNamara, "Singing to attract a mate - a stochastic dynamic game," *J. Theor. Biol.*, vol. 129, pp. 171–180, 1986.

[9] R. Howard, *Dynamic Programming and Markov Processes.* MIT Press, 1960.

[10] R. E. Bellman, *Dynamic Programming.* Princeton University Press, 1957.

[11] M. L. Puterman, *Markov Decision Processes: Discrete Stochastic Dynamic Programming*, John Wiley & Sons, 1994.

[12] H. C. Tijms, *Stochastic Modeling and Analysis: A Computational Approach*, John Wiley & Sons, 1986.

[13] R. Rezaiifar, A. M. Makowski, and S. P. Kumar, "Stochastic Control of Handoffs in Cellular Networks," *IEEE J. Select. Areas Commun.*, vol. 13, pp. 1348–1989, 1995.

[14] M. Sidi and D. Starobinski, "New call blocking versus handoff blocking in cellular networks," *Wireless Networks*, vol. 3, pp. 15–27, 1997.

[15] C.-J. Ho and C.-T. Lea, "Finding better call admission policies in wireless networks," in *Proc. VTC*, pp. 2135–2139, 1998.

[16] J. Choi, T. Kwon, Y. Choi, and M. Naghshineh, "Call admission control for multimedia services in mobile cellular networks: a Markov decision approach," in *IEEE Proceedings. Fifth IEEE Symposium on Computers and Communications, 2000, ISCC 2000*, July 2000, pp. 594 –599, 2000.

[17] Y. Xiao, C. L. P. Chen, and Y. Wang, "An optimal distributed call admission control for adaptive multimedia in wireless/mobile networks," in *IEEE Proceedings. 8th International Symposium on Modeling, Analysis and Simulation of Computer and Telecommunication Systems*, Aug. 2000, pp. 447–481, 2000.

[18] A. M. Viterbi and A. J. Viterbi, "Erlang capacity of a power controlled CDMA system," *IEEE J. Select. Areas Commun.*, vol. 11, pp. 892–900, 1993.

[19] J. Zander, S.-L. Kim, M. Almgren, and O. Queseth, *Radio Resource Management for Wireless Networks*, Artech House Publishers, 2001.

[20] R. E. Miller, *Optimization: Foundations and Applications*, Wiley-Interscience, John Wiley & Sons, 2000.

[21] A. H. M. Ross, *The CDMA Technology Site*, http://www.amug.org/ ahmrphd/, 1996.

[22] I. Karzela, *Modeling and simulating communication networks: a hands-on approach using OPNET*, Prentice Hall, NJ, Aug. 1998.

[23] S. Ross, *Introduction to Stochastic Dynamic Programming*, ACADEMIC PRESS, INC.,1993.

[24] M. L. Littman, T. L. Dean, and L. P. Kaelbling, "On the complexity of solving Markov decision problems," in *Proceedings of the Eleventh Annual Conference on Uncertainty in Artificial Intelligence (UAI–95)*, (Montreal, Québec, Canada), pp. 394–402, 1995.

[25] J. Goldsmith and M. Mundhenk, "Complexity issues in Markov decision processes," in *IEEE Conference on Computational Complexity*, pp. 272–280, 1998.

[26] N. Karmarkar, "A new polynomial-time algorithm for linear programming," *Combinatorica*, vol. 4, pp. 373–395, 1984.

[27] J. Renegar, "A polynomial-time algorithm, based on Newton's method, for linear programming," *Math. Programming*, vol. 40, pp. 59–93, 1988.

[28] C. C. Gonzaga, "An algorithm for solving linear programming problems in $O(n^3L)$ operations," *Math. Programming*, vol. 40, pp. 59–93, 1988.

[29] P. M. Vaidya, "An algorithm for linear programming which requires $O(((m + n)n^2 + (m + n)^{1.5}n)L)$," *Math. Programming*, vol. 47, pp. 175–201, 1990.

[30] K. M. Anstreicher, "Linear Programming in $O(\frac{n^3}{\ln n}L)$ Operations," *SIAM J. Optim.*, vol. 9, pp. 803–812, 1999.

Chapter 9

FUTURE TRENDS

9.1 Summary of Research

9.1.1 Wireless QoS Research

In the research from Chapter 3 to Chapter 5, we addressed the problem of call admission control and resource allocation for providing Quality of Service (QoS) to multimedia applications in the next generation wireless communication networks, which would evolve from the current wireless communication infrastructure. We considered two levels of QoS, i.e. the connection-level QoS, which describes the service connectivity and continuity under limited and varying wireless network conditions, and the application-level QoS, which measures the perceptual quality of a being-served connection by end users.

An adaptive resource management system for wireless networks to support QoS of multimedia applications was presented. The basic philosophy is to provide different QoS according to different service requests from end users under the constraint of limited and varying bandwidth resources. The system was designed to be adaptive to the network traffic conditions as well as the nature of multimedia applications. Thus, it improves user's service satisfaction in a mobile wireless environment.

On the provision of connection-level QoS measured by two metrics (the new call blocking probability and the handoff dropping probability), our service model consisted of three service classes, i.e. handoff-guaranteed, handoff-prioritized, and handoff-undeclared services, with different requirements on the handoff dropping probability. Resources were reserved in a different manner for each service class. Appropriate call admission control schemes were discussed.

Specifically, a measurement-based dynamic call admission control scheme was proposed to guarantee connection-level QoS. It also provided both connection-level and packet-level QoS for the constant-bit-rate (CBR) traffic. The scheme is flexible, computationally efficient and robust to changing network traffic conditions.

On the provision of application-level QoS, our service model considered both CBR and VBR traffics for real-time applications and the UBR traffic for non-real-time applications. For the CBR traffic, the call admission control schemes proposed for connection-level QoS provisioning can be applied to guarantee the packet-level QoS also. For the VBR traffic, dynamic call admission control scheme was developed to provide joint connection-level and packet-level QoS for the heterogeneous and varying multimedia traffic based on the analysis of the properties of the QoS metrics.

9.1.2   Resource Management Research

We addressed the problem of efficient resource management in order to provide Quality of Service (QoS) and preferential treatment to traffic with different QoS requirements. The first part for resource management research (Chapter 6) focused on the development of an efficient dynamic resource management for channel-based TDMA/FDMA systems. The second part for resource management research (Chapter 7) considered a dynamic call admission control system. The proposed system is suitable for interference-based CDMA systems by the use of interference guard margin (IGM). The third part of the research (Chapter 8) examined an optimal stationary call admission control (CAC) scheme based on the Markov decision process (MDP) model and the linear programming (LP) techniques under different handoff processes. The proposed system is suitable for channel-based as well as interference-based systems.

In Chapter 6, we presented dynamic CAC and associated resource reservation schemes based on the concept of guard channels to adapt the resource access priority by the signal-to-noise ratio (SNR) and the distance information of the potential higher-priority calls in the neighboring cells, which are likely to handoff. Under light as well as heavy traffic conditions, our CAC scheme outperformed the fixed GC scheme. The cases with different traffic profiles of mobile terminals under various traffic conditions were also discussed. We considered a comprehensive service model, which includes mobile terminals' bandwidth requirements and their different levels of priority, rate adaptivity, as well as their mobility. Our RR scheme provides more accurate estimation of potential higher-priority call arrivals, thus increasing the system reward while providing QoS guarantees to higher-priority calls. The higher system reward implies that our proposed scheme can get a good balance between resource sharing and resource reservation to achieve the opposing goals of accommodating more calls while providing QoS guarantees for high-priority class connections.

In Chapter 7, the interference-guard margin (IGM) approach has been developed by following the idea of the GC scheme. It turns out that there is a straightforward mapping from GC to IGM using the loading factor concept. A comprehensive service model was adopted. The model covered different bandwidth requirements, mobilities of mobile terminals, flexible service rates, as well as priority classes. The QoS performances in terms of several system objective functions were evaluated in the presence of various traffic characteristics. This research examined the provision of connection-level QoS from the viewpoint of the service provider (*i.e.* the system operator) as well as mobile terminals (*i.e.* users). From the perspective of the service provider, the degree of QoS is evaluated in terms of system utilization and maximum rewards. From the perspective of users, QoS is measured in terms of the new call blocking probability and the handoff dropping probability. Mathematical models for the conventional fixed GC scheme was extended to multi-threshold GC schemes to take care of the scenario with traffic of multiple priority classes. Advanced dynamic schemes were also studied by OPNET simulation. In the proposed model, we extended the analysis and simulation of traditional wireless communication networks with only one type of application (voice) to multimedia applications with different bandwidth requirements. Besides, we examined an important feature, *i.e.* rate-adaptability, for emerging multimedia compression schemes. The QoS performance was studied in the presence of rate-adaptive applications under different traffic scenarios. Based on the proposed service model and the conducted simulation analysis, we extended the preferential treatment to the 3G CDMA system. The main tool to bridge the channel-based TDMA/FDMA system and the interference-based CDMA system is the loading factor concept. It serves as a good tool to relate the received service of a mobile terminal to the amount of system resource consumption. The loading factor converts a mobile user's bandwidth, priority attribute and other characteristics, such as rate-adaptability, into a practical loading increment. Consequently, a preferential treatment can be realized in the CDMA system via the loading factor. Thus, the resource reservation scheme (*i.e.* the guard channel scheme) adopted by the TDMA/FDMA system can be applied to the CDMA system (i.e. interference guard margin scheme) in the same fashion.

In Chapter 8, an effective radio resource management scheme was developed based on the Markov decision process (MDP) model and the linear programming (LP) solution technique. They were used to find the optimal CAC policy at a given state. The proposed CAC scheme was designed for CDMA systems under a hybrid handoff scenario. The proposed stochastic control scheme determined the optimal stationary CAC policy for three traffic types (*i.e.* the new call, soft handoff, and hard handoff). From the viewpoint of the precision of CAC, the state concept in the traditional multiple threshold guard channel schemes can be viewed as a subset of the system state model discussed here. The proposed scheme has a much higher controlling precision. It was shown by computer simulation that the proposed scheme outperformed the complete sharing and the multiple threshold guard channel scheme.

9.2   Future Work

There are some interesting topics to be studied in the near future as an extension of this research.

9.2.1   Connection-level and Packet-level QoS for VBR traffic

Possible directions of future research work include the following.

- Besides the packet loss probability, more packet-level QoS measurements such as delay/delay jitter are to be considered for the joint optimization of connection-level and packet-level QoS for the VBR traffic.

- Optimal dynamic CAC schemes for various joint optimization problems with different objectives and constraints should be investigated thoroughly.

- For each proposed call admission control scheme, the robustness and sensitivity analysis under realistic wireless networks with varying and unstable conditions are interesting topics.

- The integration of the connection-level QoS service model and the application-level QoS service model shall provide a more complete framework for multimedia wireless networks. The overall system performance should be evaluated via extensive simulations.

9.2.2   Embedded System Design for Resource Management

Embedded systems have received a large amount of interest in recent years. There are some unique characteristics that distinguish the Embedded System from other computing systems [1]. First, an embedded system is often dedicated to a single function rather than general-purpose tasks. Second, an embedded system usually has tighter constraints in the cost, size, and power constraints. Third, it should process tasks and respond to actions in real time. Pagers, PDA, satellite phones, and cellular phones are some examples of embedded systems with functions described above. With the quick growth of Integrated Circuits (IC) capacity, which doubles every 18 month, it is possible to implement the resource management and call admission control functions on an embedded system.

Several types of processors can be used to realize the functionality needed in the resource management embedded system, including general purpose processors (GPPs) , single purpose processors (SPPs) , and application specific instruction-set processors (ASIPs) . The GPP has a larger size, but it often reduces the design effort and cost on required common functions. Using a GPP in an embedded system provides designer a great flexibility by working on software programming only (rather than re-design of digital circuits). The

SPP has a much faster processing speed because digital circuits are hard-wired and optimized to perform the target task. Using an SPP in an embedded system provides a better performance in terms of a faster processing speed and low power consumption, yet with little flexibility on its functionality. The third type of microprocessors are ASP, it is designed for a class of similar functionality rather than a general purpose GPP, nor too specific as SPP. Therefore, ASIP can provide the benefit of flexibility while maintaining a good performance. ASIP is a good candidate for the hardware implementation of the proposed optimal call admission control algorithm. With the advantages of flexibility and a good performance, it can be applied widely in base stations for distributed CAC and MSC for centralized CAC systems.

Figs. 9.1(a) and (b) give the I/O view of the optimal CAC cases using the Markov Decision Process (MDP) model and the linear programming technique for 2-D and 3-D systems, respectively. The input traffic parameters for the 2-D case include system capacity C, the new call arrival rate λ_n, the handoff arrival rate λ_h, the new call departure rate μ_n, and the handoff departure rate μ_h. The clock (CLK) serves for the data update and other timing purposes. The CAC module also contains N_x outputs, where each represents a call admission policy (action) for each state as described in Chapter 9.

Recently, powerful compilers have made the embedded system design feasible with high-level processor-independent languages such as C, C++. The integrated design environments (IDEs) significantly decreases the complexity of microprocessor code development and assembly language programming. This trend has enabled an easy porting of our developed algorithms to hardware design. We have the software design ready, and the gate-level design shall be carried out with the help of synthesis tools and high-level languages in the near future.

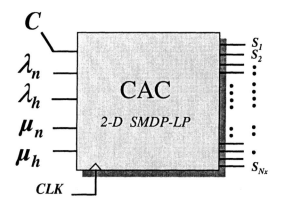

(a) I/O of 2-D CAC Embedded System.

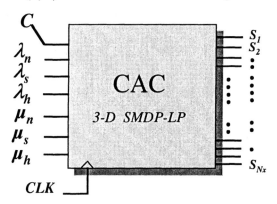

(b) I/O of 3-D CAC Embedded System.

Figure 9.1: I/O of embedded CAC systems for (a) 2-D and (b) 3-D cases.

BIBLIOGRAPHY

[1] F. Vahid and T. Givargis, *Embedded System Design: A Unified Hardware/Software Introduction*, John Wiley & Sons, 2002.

Index

1G, 9
    AMPS, 9
    NMT, 9
    TACS, 9
2.5G, 11
    GPRS, 11
2G, 6, 10
    DAMPS, 10
3G, 6, 11
3GPP, 11
3GPP2, 12

AMPS, 9
analytical modelling simulation, 163
application profile, 81
    IGM, 198
ASIP, 276

bandwidth adaptation, 48
base station
    see BS, 4
BER, 186
birth-death process, 37, 42
    N-dimensional, 42
BS, 5, 6, 16

CAC, 184
    distributed, 24
    dynamic CAC, 107
    dynamic CAC scheme
        CBR traffic, 108
        VBR traffic, 119
call admission control, 107

    see CAC, 7
call degradation probability, 50
capacity
    CDMA capacity, 186
CBR, 81, 83
CDG, 10
CDMA-2000, 12
CDPD, 4
cell, 4
channel assignment, 17
chip rate, 186
constant bit rate
    see CBR, 80
continuous-time Markov chain, 37
cost function
    IGM, 199

DAMPS, 10
DCA, 117
discrete event simulation, 163
dynamic call admission control, 74
dynamic channel assignment, 17
    see DCA, 117
dynamic guard channel scheme, 109

EDGE, 12
embedded system, 276
ETSI, 10

253